T0188341

Renewable Energy

To Godfrey Boyle, a friend, colleague and visionary,
much missed

Renewable Energy

Can it Deliver?

David Elliott

polity

The right of David Elliott to be identified as Author of this Work has been
asserted in accordance with the UK Copyright, Designs and Patents Act 1988.

First published in 2020 by Polity Press

Polity Press
65 Bridge Street
Cambridge CB2 1UR, UK

Polity Press
101 Station Landing
Suite 300
Medford, MA 02155, USA

ISBN-13: 978-1-5095-4163-8
ISBN-13: 978-1-5095-4164-5 (pb)

A catalogue record for this book is available from the British Library.

Library of Congress Cataloging-in-Publication Data

Names: Elliott, David, 1943- author.
Title: Renewable energy : can it deliver? / David Elliott.
Description: Cambridge, UK ; Medford, MA, USA : Polity, 2020. | Includes
    bibliographical references and index. | Summary: "An incisive introduction
    to renewable energy and its global expansion"-- Provided by publisher.
Identifiers: LCCN 2019052504 (print) | LCCN 2019052505 (ebook) | ISBN
    9781509541638 | ISBN 9781509541645 (pb) | ISBN 9781509541652 (epub)
Subjects: LCSH: Renewable energy sources.
Classification: LCC TJ808 .E453 2020 (print) | LCC TJ808 (ebook) | DDC
    333.79/4--dc23
LC record available at https://lccn.loc.gov/2019052504
LC ebook record available at https://lccn.loc.gov/2019052505

Typeset in 10.5 on 12pt Sabon
by Fakenham Prepress Solutions, Fakenham, Norfolk NR21 8NL
Printed and bound in the UK by TJ International Limited

For further information on Polity, visit our website: politybooks.com

Contents

Acknowledgements

Thanks for some criticism, advice and input from Tam Dougan, editor of the long-running *Renew* newsletter, previously based at the Open University and now run independently.

Boxes

Abbreviations

BECCS	bioenergy with carbon capture and storage
BECCU	bioenergy with carbon capture and utilization
CCS	carbon capture and storage
CCU	carbon capture and utilization
CHP	combined heat and power
CO_2	carbon dioxide
CSP	concentrated solar power
DACS	direct air capture and storage
EROEI	energy return on energy invested
GDP	gross domestic product
IEA	International Energy Agency
IRENA	International Renewable Energy Agency
LNG	liquid natural gas
LUT/EWG	Lappeenranta University of Technology/Energy Watch Group
NET	negative emissions technology
NGO	non-governmental organization
P2G	power-to-gas
PV	photovoltaic solar power
REN21	Renewable Energy Network for the 21st Century
WEC	World Energy Council

Preface

The use of renewable energy is spreading rapidly, and some claim that wind, solar and other renewables can and will become the dominant global energy sources within a few decades, thus avoiding major climate change problems. Concerns about climate change have been a key driver, leading to growing government support. However, the falling cost of renewable energy has also become a major driver. Some renewable energy technologies are now competitive across the board and costs continue to fall, with a new commercial dynamic adding impetus to their uptake. This book asks whether that trend will be sufficient to ensure that renewables expand fast enough globally to limit climate change to survivable levels without imposing high costs.

There are certainly many who doubt that this is possible. Some critics argue that renewable energy systems are inherently unreliable and expensive, and they look to nuclear power as an alternative non-fossil option, and possibly also to cleaning up fossil fuel use. This book seeks to meet these claims head on and asks to what extent renewables can deliver a technologically and economically viable energy future, and whether other technical options and energy policies to support them are also needed. It explores how important renewable energy technology might be by looking at its progress so far and at its future potential and problems in a context where other approaches are also on offer. Much has been promised from renewables and, so far, they seem to be living up to the promises as they accelerate ahead. This book looks at whether that can and will continue.

1
Introduction: All Change?

Renewables are no longer marginal but have become mainstream. This introductory chapter asks what has driven their recent success – and can it continue?

The dynamics of technological change

The way in which we have used technology to meet energy needs in the past has been a key cause of many environmental problems, including air pollution and climate change, but new technology, and new patterns of development based on it, may offer possible solutions to some of these problems. That said, while there is a strong case for looking at technology as a key factor in the attempt to move towards a more environmentally sustainable global future, technology is only *one* factor and possibly not the leading one. We may also need social and economic change.

However, there are interactive processes at work. The development and adoption of new technology is usually driven by social and economic forces, including profit. Environmental issues have also increasingly come to the fore and are having a direct effect. For example, climate concerns have been a stimulus for rapid recent renewable energy growth, along with air pollution issues, notably bad of late in some Asian cities. Something had to be done, and a switch to cleaner technologies was one option.

Some of these changes have been led by governments, nationally and internationally. In addition to supporting global climate actions designed to reduce carbon emissions, most countries have backed the rapid expansion of renewables, with more than 50 countries having signed up to '100% by 2050' renewable power targets (REN21 2018). Clearly, concerns about climate change have translated into policy changes and action programmes (see Box 1.1 for an overview of the key energy-related climate issues).

Box 1.1 Energy use and climate issues – an overview

Around 80% of the energy used globally comes from roughly equal amounts of coal, oil and natural gas, hydrocarbon materials which were laid down in geological fossil strata eons ago. We are about halfway through extracting and burning off the easily available fossil fuels, but it seems unlikely that we can burn off the rest without causing major environmental and social problems. A key issue is that the combustion of coal in power stations, gas for heating in homes and oil products in vehicles, along with other activities like cement making, produces carbon dioxide (CO_2), a so-called 'greenhouse' gas that, rising into the upper atmosphere, acts like the windowpanes of a greenhouse, trapping incident solar heat inside. So the greenhouse – in this case the earth – heats up. We are headed for maybe a 4–5 degrees Celsius (°C) initial average global temperature rise over the next century, and possibly more if the polar region permafrost zones thaw out, releasing trapped methane gas, a much more potent greenhouse gas than carbon dioxide. Even without that, the climate models suggest that we are likely to face progressively worsening threats to the already badly (pollution-) stressed environment and ultimately perhaps to ecosystem stability and survival (IPCC 2019).

Current global climate policy, as agreed in Paris in 2016, looks to holding the temperature rise below 2°C, and ideally below 1.5°C, but even holding it to 2°C may not be possible. As the weather/climate system changes and becomes more erratic, we are therefore likely to experience more major storms, floods, droughts and worsening wildfires and thermal stress. As the icecaps melt and the seas warm and expand in volume,

> sea levels will rise, threatening many coastal cities and food-growing areas. It may be too late to avoid some of this, so we have to be prepared, but there is also an overwhelming case for taking urgent action to reduce carbon dioxide (and methane) emissions so as to avoid it all getting very much worse (Carbon Brief 2018).

To some extent, while, in part, a response to physical threats to the environment and to human health, the political actions being taken around the world are also based on assessments of the likely *economic* impacts of climate change and pollution. It has been argued that the economic cost of inaction would vastly outweigh the cost of responding to the threat by maybe a factor of ten or more (Stern 2007).

This type of argument laid the base for much that has happened until relatively recently. It might be costly, but a change had to be made. The main policy issue was therefore how the cost of 'decarbonizing' could be met, with there being no shortage of political resistance to proposals that would, it was assumed, increase energy costs. That was in a context where the threat of climate change was sometimes seen as longer term, whereas there were arguably more pressing short-term economic and political concerns. So there was some resistance.

Renewable energy, often promoted as the main way forward, was seen as expensive, even more so than the nuclear alternative, usually seen as its main rival, so that the cost of the change might be prohibitive and politically difficult to impose. However, there have been new developments which may change the situation. While nuclear costs remain high, and even seem to be rising (WNISR 2019), it may now be the case that the newly emerging renewable energy technologies can be taken up without imposing extra costs. They may even deliver a system that is actually more economically viable than the present one. Moreover, renewables are now getting so cheap that their uptake may accelerate under market pressures, regardless of whether climate change or pollution is taken seriously (PEI 2018).

That opens up some interesting possibilities. Although the environmental and climate benefits are still obviously important, some renewable energy proponents have suggested that renewables no longer need to use climate change as a justification. They can now prosper on their own economic merits. Indeed, for many countries, it is claimed, the direct economic benefits, and their role in fuel-cost saving, have become, or will become, the main motivation for investment in renewables. Thus it has been argued that 'renewable generation will stand on its own commercial feet and cutting down emissions will become a fortuitous side-effect' (Swift-Hook 2016).

That may be going too far, but it is certainly clear that the cost of renewables has fallen dramatically, in particular for PV solar (Lazard 2018a). According to the International Monetary Fund, 'between 2009 and 2017, prices of solar photovoltaics and onshore wind turbines fell most rapidly, dropping by 76% and 34%, respectively – making these energy sources competitive alternatives to fossil fuels and more traditional low-carbon sources' (IMF 2019).

That trend also extends to offshore wind, initially seen as one of the most expensive renewable energy options, with 'strike prices' under the first round of the UK's CfD capacity auction system in 2015 reaching nearly £120/MWh. In the 2019 CfD round, strike prices for some successful project bids fell to just under £40/MWh, around a third of the earlier figure (New Power 2019).

These cost falls have been important in propelling renewable energy to the fore, and they are likely to continue. So it has become very hard for opponents to maintain resistance to what seems likely to be an unstoppable change dynamic. However, they do sometimes try. For example, it is ironic that the incumbent fossil and nuclear interests, which often initially dismissed the renewable energy options as totally irrelevant, now have to fight to protect their market shares as opposition to fossil fuel use mounts, nuclear costs rise (Portugal-Pereira et al. 2018) and renewables win out across the board. They already supply more than 26% of global electricity (REN21 2019), compared to nuclear power at around 10%, and are displacing fossil fuels in many markets.

Interestingly, under increasing pressure from renewables, some of the fossil and nuclear lobbies have therefore changed tack a little and are now arguing that renewables are still problematic since they cannot expand *fast enough* to deal with climate change (BP 2018a), with some climate change sceptics joining in (Lyman 2019), this after they have all resisted change and spent so long trying to stop renewables from getting started. Some of the incumbents may continue to resist, but over 185 companies so far have signed up to '100% by 2050' renewable electricity targets (RE100 2019).

As argued above, what has changed things is not a sudden extra concern about climate or air pollution (although, as noted, that has happened and has helped) or even what some see as a collapse of nuclear power as a future option (WNISR 2019) but the fact that the cost of renewables has fallen dramatically. Arguably, a new economic dynamic has, at least partially, taken over, with renewables well placed to become the dominant option.

However, while the case for renewables does look strong, they are up against a set of well-established energy technologies, well entrenched in lucrative markets. The fossil fuel-based incumbents look to new carbon capture technology to allow them to stay in the game, and the nuclear lobby similarly looks to new technology to improve its economics. To set the scene, Box 1.2 provides a short summary of the main impacts of the options on offer, and I will be looking at the options, and at the issues raised by them, in more detail later.

Box 1.2 The new energy options – a summary of impacts and issues

The use of naturally and continuously replenished *renewable energy flows*, like the winds, waves, tides and solar heat/light, produces no direct carbon dioxide (CO_2) or other emissions. There will be *indirect* emissions due to the use of fossil fuel for the *construction* of the technologies and for the production of associated materials, but that is true, at present, for all energy technologies. Once built, renewable energy-based power plants, like wind turbines and solar farms, differ from the rest in not needing any fuel to run. However, they may have some local impacts, and some (but not all) produce variable power outputs.

Those issues apart, they are strong 'clean energy' contenders, arguably more so than nuclear plants, which, although they do not produce CO_2 in operation, rely on the use of fossil fuel to mine and process/enrich their fuel, a very energy-intensive process, thus incurring a carbon debt. There are also long-lived radioactive wastes to deal with, as well as the risk of leaks and unplanned release of radioactive material. Global fissile fuel reserves are also finite; they are not a renewed resource. Nuclear fusion, as opposed to fission, is still some way off as a practical option and may remain so but might have fewer fuel resource limitations, although there could still be risks and radiological implications.

It is possible to capture and store the CO_2 produced by fossil fuel combustion plants, but, although that might allow us to continue to use fossil fuel, as I will be describing, there are operational and economic limitations to this arguably rather inelegant 'end of pipe' engineering approach to post-combustion 'carbon capture and storage'. The environmental argument is that we should not be burning fossil fuel in the first place nor trying to find places to store the resultant CO_2 safely and indefinitely. The global fossil fuel resource is in any case finite, so using it is not a long-term option, even ignoring CO_2 and other emissions and impacts, for example in relation to air quality.

The combustion of biomass (plants, wood and other bio-materials), and then the capture and storage of the CO_2 produced, is an option. In theory, since CO_2 is absorbed when biomass is grown, that process would be carbon negative, reducing net atmospheric CO_2 levels. However, to have a significant CO_2 impact, in addition to vast CO_2 storage requirements, very large amounts of biomass would have to be grown and burnt, with large land-use and ecological impacts.

The capture and storage of CO_2 direct from the air is also possible, although that process would *use* energy rather than generate it. As an alternative, some of this CO_2, or the CO_2 from power plants, might be used to make new hydro-carbon fuels, if a source of hydrogen were available, for example produced using renewable energy. However, burning the resultant synthetic hydrocarbon fuels would release the CO_2 again. It might be better to use the 'green' hydrogen direct as a fuel since its combustion only produces water vapour.

I will be coming back to these various options, issues and choices later, for example looking at costs, but from this short summary it does look as though, in terms of clean energy supply, renewables have the edge environmentally.

Can renewables deliver?

While in general terms the prospects for the future of renewables may look positive, and the overall case for alternatives may look poor, the resistance of incumbents, and some of the arguments against renewables that they have adopted, do have to be faced. A central issue raised is the question of whether renewables can expand rapidly enough to meet global energy needs.

This was met head on in a scenario published initially in 2009 in *Scientific American* (Jacobson and Delucchi 2009) and then more formally in 2011 (Jacobson and Delucchi 2011) and developed in their subsequent studies. It was suggested that a global target of obtaining 100% of all energy from renewables by 2050 was viable, at reasonable cost. That was ridiculed by critics as impossible, and there were debates over methodology (Clack et al. 2017). However, now, with several countries already above 50% and many dozens of further studies from around the world suggesting that very high renewables shares are possible (Stanford 2019), the debate is more about total-system costs and whether it will 'only' be 70%, or more than 80% (of electricity), globally by 2050 (IRENA 2017a), or how to do better than that (Bogdanov et al. 2019).

The pace of technical development and market adoption has been startling, taking even enthusiasts by surprise (for photovoltaic solar especially), and can be contrasted with the slow pace of development of the rival technologies, nuclear (Pearce 2017) and carbon capture and storage (Simon 2017).

As the head of the UK government's advisory Committee on Climate Change put it, they had initially been 'overly optimistic about cost falls in some other technologies – nuclear for example', but for renewables 'innovation has been the key – driven by policy – in ways that we did not fully expect ten years ago. Globally, a clear goal to decarbonise, with co-benefits of improved air quality in cities, has stimulated commercial innovation' (Stark 2019a).

However, there are inevitably issues with renewables. Some of them have been highlighted in recent critiques from, amongst others, pro-nuclear lobbyist Michael Shellenberger.

He says that renewables cannot power modern civilization, given that the energy sources are variable and also dilute and diffuse, requiring the use of large areas and involving significant local environmental impacts, as well as risks to human and animal life, along with high costs for backup requirements (Shellenberger 2019). He is not alone in challenging the viability of renewables. There is a range of critical books, articles and reports at varying levels of coherence (Lomborg 2019; Montford 2019; Rogers 2018).

It is relatively easy to provide specific counters to these challenges and to the assertion often made that 'nuclear is a better bet than renewables'. For example, on cost, it is clear that many existing nuclear plants in the United States are having to close because they are no longer competitive (Abdulla 2018) and that, globally, few new nuclear projects are going ahead. By contrast, renewables are winning out economically in most countries (WNISR 2019).

On safety, so far globally there have been around 192 people killed in accidents related to wind farms, mostly involving occupational accidents during installation or maintenance work, but some involving blade transport (CWIF 2019). By contrast, estimates for deaths associated just with the 1986 Chernobyl nuclear accident, although debated, range up into the thousands and possibly tens of thousands (Ritchie 2017). It is true that emissions from coal-fired plants lead to many more deaths, for example from respiratory illnesses, quite apart from any climate change-related impacts, but arguably the solution is to go for renewables like wind and solar, not nuclear, as an alternative.

The wider environmental impact issue is a bit more complex, as I have explored in detail elsewhere (Elliott 2019a). In general, renewables are land using, some more than others (e.g. PV solar can be on rooftops, offshore wind uses no land), but the complete nuclear fuel cycle, from uranium mining to waste deposition, also involves significant land use, with power plants, fuel-processing facilities and so on having to be protected by large fenced-off areas for security and safety. Nuclear plants do not generate CO_2 gas directly themselves, but, as noted in Box 1.2, producing the fuel for them is a very energy-intensive process, mostly at present based on the use of fossil fuels.

By contrast, renewables like wind and solar do not need any fossil fuel to run, so they are totally carbon free in operational terms.

While most renewables have generally low, or even negligible, global environmental/climate impacts, some can have significant local impacts, large hydro in particular. For most other renewables (including small hydro), there are technical options that can reduce local operational impacts on wildlife, such as acoustic bird-scaring systems for wind turbines, and there are also ways to avoid or reduce local social impacts by careful design, siting and operation. Although there are areas of marginal land that can be used, biomass is probably the worst offender in terms of land use. Growing biomass energy crops is inevitably land using. That is one reason why there is now more interest in using biomass in the form of farm and food wastes since that already exists: using it can be part of a move to a lower-impact circular economy. In terms of climate impacts, since the CO_2 produced when biomass is burnt is re-absorbed when plants grow, biomass can in theory be near carbon balanced *if* the rate of use is matched by the rate of replanting. Nevertheless, as I will be illustrating, although it can be a *renewed* resource, the use of biomass as an energy source may have eco-impacts, depending on the type of biomass and its pattern of production and use.

There is also the obvious, more general point that there is a need to balance the *variable outputs* from renewables like wind and solar, the cost of this often being presented as a 'killer argument' against them. However, as I will be exploring in detail in subsequent chapters, it is not an insuperable problem. The grid system already handles variations in supply and demand and can be upgraded to continue to do that as more renewables are added, although it may take time to develop and deploy some of the new technologies that will be needed, including storage capacity. The extra cost of grid balancing has been put at maybe 10–15%, or perhaps less, if the proper measures are adopted (Heptonstall, Gross and Steiner 2017): some of the new grid-balancing measures may *reduce* system costs by matching energy supply and demand better, thus improving overall system efficiency (ICL/Ovo 2018).

A more substantial issue is that there will be a need to supply heat and transport energy as well as electricity, a somewhat harder task. Nevertheless, as I explain below, it can be done, although to understand how, and to get to grips with the full transition costs, we need to start looking at the energy system as a whole, not at individual components. Making system-wide changes may be hard and, although the commercial incentives to move ahead are now stronger, they may not be sufficient to accelerate renewable expansion and system change fast enough to deal with the urgent climate and pollution problems. So there may be a need for extra support from governments, for example via subsidies to enable accelerated programmes of development and deployment. That has certainly been the lesson so far: markets on their own have not been sufficient. The point may soon be reached when subsidies will no longer be needed, at least for the initial wave of renewables, but clearly there will be cost implications and political choices associated with making the energy transition (Carbon Tracker 2019). They are what this book aims to explore.

Policy change – the costs of the transition

In the short term, some say, the changeover to renewables may incur extra costs, perhaps, according to one EU-focused study, adding up to 30% to the total-system cost (Zappa, Junginger and den Broek 2019). However, that view has been disputed in the case of the European Union (EU) (Beam 2019) and is also challenged in many of the global 'high renewables' scenarios that have emerged. Instead, it is argued that, as the new system develops, direct and indirect costs should fall since there would be no use of increasingly expensive fossil fuels, and the social and environmental costs of their use would be avoided. As renewable costs continue to fall, and climate threats rise, that view does seem attractive.

Nevertheless, a big political issue *in the interim* is whether the falling cost of renewables will ensure that the cost of their rapid expansion will avoid a backlash from consumers. The implementation/support costs certainly have been a problem. As a result, some expansion programme slowdowns have been imposed (across the EU and also in China), ostensibly on the basis of fears about the rising costs of support schemes. These cutbacks have been buttressed by the rise of populist backlashes from those who feel they have been left out or left behind in social and economic terms, which has added a new political dimension and feeding back, around the world, into reaction against ostensibly progressive change (O'Neill 2018). In parallel, in relation to the cost of energy transitions there have been warnings, from otherwise divergent camps, about the need for 'energy justice' (Monyei et al. 2018) and the risk of increased fuel poverty (Beisner 2019).

While there are certainly social equity issues to be addressed (McGee and Greiner 2019; Sovacool 2013), and I will look at them later, there are also wider strategic energy perspectives which may present challenges to renewables, such as the belief that other options would be better and cheaper, for example nuclear and fossil carbon capture (Aris 2018).

Technically, the renewables case is strong. The International Renewable Energy Agency (IRENA) claims that renewable energy, along with energy efficiency, can provide more than 90% of the necessary energy-related CO_2 emissions reductions (IRENA 2018). The economic impact case is also good: despite fears about the cost of the energy transition, it may not in fact cost consumers too much. The European Commission (EC) says that, under its proposed renewables-led transition, 'by 2050, households would spend 5.6% of income on energy-related expenses, i.e. nearly 2 percentage points lower than in 2015 and lower than the share in 2005' (EC 2018). However, that is speculative, and the EC does include more than just renewables in its proposed mix, as do some other studies. Indeed, although most agree that renewables will boom, some see fossil fuel, and possibly nuclear, remaining as a vital part of the energy mix into the far future (BP 2019; WEC 2019).

So the question remains: are they right, or can the renewables meet *all* our energy needs? That may depend on what sort of future global energy system and economy we are looking to create, how rapidly the changes can be made and how we go about making them, issues explored in subsequent chapters.

The first part of the book focuses on the technological aspects, starting off, in chapter 2, with an overview of the technological options and key related transition issues. Chapter 3 then looks at what energy supply technology choices have been made in some of the existing future-energy scenarios, while chapter 4 looks at system integration and balancing requirements and options.

The second part of the book moves on from essentially technologically defined issues and options to an exploration of the wider strategic issues and choices, including social change options, asking whether we can and should move to a low- or zero-growth sustainable energy future. Chapter 5 looks at the issue of growth, and at the need for a socially equitable transition process, while chapter 6 focuses on the wider geopolitics of the transition. Chapter 7 looks at some examples of what is happening around the world, leading to an overall conclusion in chapter 8.

Energy metrics and climate impacts – a short guide

The emphasis in this book is on energy policy, in particular the technological choices ahead. I have tried to use plain English and avoid technical terms but, inevitably, getting on top of what is a complex field requires some understanding of basic energy systems and technology and related issues. Box 1.3 provides a short guide to the *measurement units* used in this book. Perhaps surprisingly, not all of the metrics used are uncontroversial.

Box 1.3 Energy and power units

The terms 'power' and 'energy' are sometimes used inter-changeably, which can be confusing. In this book, 'power' is used to mean electric power, whereas 'energy' covers all sources/end uses (power, heat and transport), not just electricity, although of course some electricity is used for heating and for transport.

In physics, the definition of *power* is more specific. It is a measure of the ability (or capacity) of a device to do work, rendered in watts and multiples of watts: 1,000 watts is a kilowatt (kW), 1,000 kW is a megawatt (MW), 1,000 MW is a gigawatt (GW), 1,000 GW is a terawatt (TW). *Energy* also has a more specific meaning in physics. Strictly, it is always conserved and cannot be 'generated' or 'consumed', only converted from one form to another. But we still commonly talk of energy generation and consumption, and the amount of energy converted (generated or consumed) is measured as power × time, i.e. the power of the device multiplied by how long it is run for, so it is 'kilowatt hours' (kWh) and multiples, MWh, GWh and TWh.

There will always be *losses* in energy conversion from one form to another and so, for a generation system, the finally available end-use energy will be less than the so-called *primary* energy inputs (for example, the energy in the fuel used in fossil-fuel-fired plants). The actual energy output of a generation plant will also usually be less than the theoretical full output possible for the plant, especially for systems using variable renewable sources, since they cannot deliver their full theoretically possible output all the time. 'Capacity factors' (also called 'load factors') are cited for the percentage of the theoretical maximum output capacity that is actually available annually to meet demand loads.

There are some issues with the way renewables are handled in energy analysis, since renewables like wind and solar do not use fuel. To produce a figure for *primary energy* that is compatible with those used for fossil fuel plans, the output from the renewable plant is sometime 'grossed up' by a factor of around three, to reflect the amount of primary fossil energy that would have to be used (given the large losses in fossil energy conversion) to produce the same output. The same is sometimes done with nuclear plants and biomass plants. It might be argued that it would better just to compare final energy outputs in each case. But, done that way round, to get the same output as a wind or solar plant a fossil plant could be depicted as having, nominally, to consume around three times more primary fuel (Sauar 2017).

Although, as I will be exploring, there can be disagreements about the likely technical and economic viability of new energy technologies, the wider climate change and environmental aspects and issues are sometimes even more controversial. For example, there are sometimes fierce debates concerning the likely social and environmental impacts of the combustion of fossil fuel and energy use, and also over how they might be avoided or lessened. However, the basic emissions situation is relatively clear. Box 1.4 summarizes the current global pattern of energy use, in very simplified terms, and also, in rough outline, the resultant global carbon emissions. They are both still expanding.

Box 1.4 Energy end uses and emissions

In terms of what the various energy sources are used for, put very simply, although it varies significantly around the world, the *total primary energy* that is used for (electric) power generation, for heating and for transport is very roughly split in equal amounts amongst these three end uses, but that pattern is changing with, in some cases, transport taking more.

In terms of *carbon dioxide* gas emissions from fossil fuel use, total emissions from direct energy production/use have stabilized globally in recent years but rose slightly (by 1.7%) in 2018. Although it varies round the world, there are, very roughly, equal proportions of global greenhouse gas emissions from energy generation (heat and power), transport, industry and agriculture (IEA 2019a). The historical record of emissions illustrates how emissions rose as countries industrialized, led initially by the United Kingdom and then the United States but with China now in the lead (GCP 2020).

For details of energy use and some of the resultant impacts, the UK situation is reported in the annual Digest of UK Energy Statistics (DUKES 2019), the Energy Information Administration produces data on the US situation (EIA 2019), while BP publishes annual global energy outlooks (BP 2019), as does the IEA (IEA 2019a). The IEA's latest data suggest that global emissions have stabilized again, in part due to reductions in coal use in the United States and the EU (IEA 2020).

The precise extent and timing of the likely social and environmental impacts from the combustion of fossil fuel may still be the subject of some debate, but few doubt that they will continue to grow and cause serious and worsening problems unless major changes are made. Indeed, some fear that, otherwise, the outcome could be catastrophic. Whatever we now do to slow emissions, the impacts from past emissions will have to be faced, in some cases as a matter of urgency. We can also try to adapt in advance to climate change to some extent via ameliorative measures, such as enhanced sea-rise protection in flood-prone areas.

However, if nothing is done to halt or reduce emissions, the impacts will get worse, making adaptation progressively harder, more expensive and in the end futile. It is the same for carbon dioxide removal from the atmosphere: at best, it can deal with some old emissions but more likely, as with post-combustion carbon capture and storage of CO_2 from power plants, it may just be used to compensate for continued fossil fuel use. Carbon capture nevertheless might reduce the associated climate impacts, although there will be limits to the storage space for CO_2. As with adaptation, it is not a long-term answer to climate change and the impacts of continued fossil fuel use. Avoiding emissions at source is a more fundamental, effective and sustainable approach (Schumacher 2019a).

My aim in this book is to ask, how far can the use of renewable energy sources allow us to move in that direction? Can they help us to cut emissions substantially or even *entirely*, and, if so, when and at what cost?

2
The Renewable Transition

To set the scene, this chapter looks in outline at the key renewable options and systems, their potentials, costs and problems. It reviews the basic issues of choice at stake and also looks at how rapidly the options might be deployed.

The renewable options

Natural energy flows can be tapped and converted into mechanical power and then electrical energy, as in hydro projects, and by wind-, wave- and tide-driven devices. In addition, there are systems which use natural sources of heat, either directly for heating or indirectly to generate electrical energy, and finally there are devices which convert solar energy directly into electricity, using photovoltaic cells (PV solar).

I have explored these options in some detail in a recent book (Elliott 2019a), so here I will simply present a brief summary of the state of play. The most developed option so far is hydroelectric power, with around 1.2 terawatts (TW) of capacity installed globally, supplying about 16% of the world's electricity, producing it relatively cheaply. Indeed, although they are expensive to build, once in place hydro plants often offer some of the lowest-cost electricity on the grid in many countries.

However, hydro has struggled with negative environmental and social impact assessments, and with occasional dam failures and loss of life. For example, the failure of the Banqiao Dam in China in 1975 led to 26,000 deaths from flooding. In some cases, hydro has also been beset by unreliable rainwater supply due to climate change (Moran et al. 2018). As I will be considering later, smaller projects, including run-of-the-river schemes without reservoirs, are often favoured by environmentalists since they may have lower local impacts, but hydro projects with large reservoirs can play an important 'pumped storage' role in grid balancing. In such schemes, surplus green power from wind or PV solar projects is used to pump water uphill into the reservoir, ready for generation when needed to meet lulls in the availability of power from the other renewable sources.

The same thing may be done with large cross-estuary tidal barrages. Tidal barrages capture a head of water behind a dam at high tide and this (along with any extra head created by pumped storage) can be used to generate electricity, as with hydro plants. Smaller artificial tidal lagoons, impounding areas of open water, are another option, and they too can be run in pumped storage mode. Although both these tidal-power generation systems are technically viable, and two medium-scale (250 MW) barrages have been built, large tidal barrages are expensive and are usually opposed by environmentalists as being too invasive (they block off entire estuaries), while as yet no tidal lagoons are in operation.

These so-called 'tidal range' systems operate on the vertical rise and fall of the tides. By contrast, tidal current turbines (in effect, underwater wind turbines) operate on horizontal ebbs and flows and have proved to be a more popular and successful tidal option. They have also been easier to develop than the other potentially large sea-power option, wave energy. It is harder to extract energy from waves, and from the often chaotic interface between water and wind on the surface, than from the smooth tidal flows beneath it. Nevertheless, many wave projects, as well as tidal current projects, are under development (JRC 2016) and, although costs for tidal turbines and wave devices are still quite high, we might expect them to fall, as has happened with offshore wind. So there may be GWs deployed soon, possibly 1 TW or more eventually, with low environmental impact.

While these new options are still mostly at the development stage, wind power, on- and offshore, has been the big new technology success, with costs falling dramatically and global capacity heading for 600 GW and 1 TW soon (IRENA 2019a). The rate of expansion in many parts of the world is staggering, in China especially but also in parts of the United States and much of Europe. Longer term, 5–10 TW or more may be possible, including increasing amounts offshore.

It may be a wild card but the wind resource could be dramatically expanded if *airborne* devices prove to work well and safely. Mounting turbines on autogyro-like flying devices or giant drones, delivering electricity to the ground by cable or using the pull from tethered kites to drive ground-mounted generators is some way off, and there are many issues. However, systems like this, perhaps used in remote offshore areas away from aviation corridors, would give access to the much larger wind resource higher up in the atmosphere (Bown 2019).

That is all very speculative. For the present, large *floating* offshore wind devices are the main breakthrough technology, able to operate in deep water far out to sea, where direct seabed-mounted supports would be prohibitively expensive or impossible. Systems are under test in the EU and Asia. Like offshore wind in general, they avoid the land-use and visual intrusion issues that have sometimes constrained onshore wind projects (Equinor 2019).

Biomass, in its modern usage for electricity production, has faced even tougher land-use and eco-impact constraints, particularly in relation to the impact of the use of forest-derived biomass on carbon balances and carbon sinks: replacement growth can take time. There have also been major concerns over the land-use and eco-impacts of vehicle biofuel production. As I will be describing, views clearly differ on whether biomass can be relied on as a major source of heat, electricity and transport fuel, but, as noted in chapter 1, some look to the use of bio-wastes to avoid the eco-problems and extra land use.

The International Energy Agency (IEA) says that global biomass electricity-generation capacity (including that using bio-wastes) is anticipated to reach 158 GW by 2023, and it also looks to expansion of biomass use for heating and vehicle fuel production (IEA 2018a). Given the environmental issues, that may be optimistic or even undesirable, but Fatih Birol, the IEA's executive director, has said that 'Modern bioenergy is the overlooked giant of the renewable energy field. Its share in the world's total renewables consumption is about 50% today, in other words as much as hydro, wind, solar and all other renewables combined' (Chestney 2018).

Direct solar heat use, for example via solar heat collectors on rooftops, or large ground-mounted arrays, has had far fewer problems and is heading for 500 GW (thermal) globally, with, in some cases, large heat stores offering a way to use summer solar heat in the winter, although at a price. Concentrated solar power (CSP) conversion of solar heat to power also has a large potential. It uses sun-tracking mirrors or parabolic focusing dishes or troughs to raise steam or other working vapours to drive a turbine generator. However, CSP has been less successful than PV solar so far, with only around 5.5 GW of CSP in use globally. That is despite the fact that CSP has a heat storage option (using tanks of molten salt), so that the power generators can run at night, enabling the plant to operate 24/7, unlike PV solar. This is a significant advantage, enabling CSP to deliver firm, continuous power, but it comes at a cost and with limits. Unlike non-focused PV, CSP needs direct, as opposed to diffuse, sunlight, so it is mostly used in desert areas and, although arid, deserts do have fragile ecosystems which CSP can disturb.

There is also the solar chimney option. Solar heat is collected in greenhouses in hot desert areas, with the hot air fed to an updraft chimney, driving an internal wind turbine as it rises. Small- to medium-scale prototypes have been tested in Spain, China and elsewhere but to get efficient power output the towers would have to be very tall, several hundred metres. Ocean thermal (OTEC) solar devices, which work on local temperature differentials between the surface and the depth of the seas, in effect using the sea as a vast solar heat collector, also hold some promise, but it is a very location-specific option, most relevant to warm sea areas in the Pacific. Geothermal energy from deep under ground is also location-specific but is making progress (13 GW power, 28 GW heat, so far), and the long-term global power potential is large, 2–3 TWs or more with, once the wells have been established, the local impacts being low (IRENA 2017b).

The big success story, though, is solar PV, now heading for 500 GW globally, with costs falling very rapidly, so much so that some look to mass, multi-TW deployment in the years ahead, maybe as much as 20 TW or, as explored later, even more by 2050. Some of this is due to new technology, including new, more efficient high-tech multi-junction cells with new materials, some of them, with light focusing, getting to 40% or more energy-conversion efficiency.

Even with lower efficiencies than that, more conventional PV arrays have spread around the world, including some very large projects (some over 1 GW) in desert areas but also many smaller arrays (typically of up to 20 MW) in rural solar farms, as well as many millions of individual units (of a few kW) on domestic and other rooftops.

While efficiency is important, so is cost, and there are now cheap, easier to mass-produce thin-film or dye-based cells, increasingly using non-toxic materials. They have low efficiencies so far, but they can be used for many new applications, including solar windows, with flexible, spray-on PV materials opening up many new deployment options. PV of various types is already being used for solar roads, solar car-port canopies and, increasingly, in floating arrays, for example on reservoirs (Chandran 2019). That helps deal with one of the big drawbacks of PV: once you have used all the rooftop space available, it takes up land space.

An alternative is to put PV arrays *in* space, in geostationary orbits, beaming energy to earth by focused microwave links (Snowden 2019). Although very costly, and technically difficult, that would deal with the other big drawback of using solar: it gets dark on one side of the planet at night. While location in permanent sunshine in space is tantalizing, possibly offering solar supply to sites on earth at night, there may be cheaper ways to achieve round-the-clock supply than by launching PV into space and microwaving power back. New storage options are now emerging, along with new long-distance supergrid transmission possibilities. It may, in time, even be possible to shunt power from the sunny side to the night side of the planet, using terrestrial interlinks between grid systems, the 'global grid' idea, which I will be looking at later (see chapter 4).

Renewable balancing

That brings us to grid balancing and system integration, an area that is expanding in importance as the use of variable-output renewables increases. There will be times when there are local or regional lulls in energy available from the wind and the sun and, although some of the other renewable sources are not variable, there will be a need at times to provide balancing grid power (Elliott 2016).

One view sometime expressed by opponents of renewables is that *each* renewable energy project will have to be backed up 100% by conventional capacity. A moment's thought should indicate that this is not the case. All power plants, of whatever type, can have downtimes, and we do not insist that each has its own individual backup. Instead, it is the *power system as a whole*, including all the other power plants, that provides backup, and it is the same for renewables. They may however need backup inputs *more often* than conventional plants.

Initially, with relatively low levels of renewable input, much of this extra balancing input can be provided by the remaining conventional plants, for example gas turbines, so we will not need new capacity – it already exists. It is the main way in which supply and demand variations are balanced on the grid. With variable renewables on the grid, the gas plants have to ramp up and down to full power a few times more. Some of that capacity can be gradually changed over to using non-fossil green gas so as to avoid emissions but, as renewables expand, extra balancing capacity may also be needed.

A simple view is that energy storage will solve everything, especially as it is getting cheaper. Sadly, it is more complex than that (Elliott 2017a). Batteries are getting cheaper, but they have low storage capacities and are best used to store electricity for a short time, a few hours or days at most. They can also be used to deal with short-term voltage and frequency perturbations on the grid, but for longer-term variations, and long lulls in power availability, you need large higher-capacity bulk-storage systems. Pumped hydro plants can perhaps cope for a few days, if they have large reservoirs. For longer than that, you need something extra.

Options for longer-term bulk storage include compressed air and hydrogen gas, stored in vast underground salt caverns. The hydrogen can be made by the electrolysis of water, using surplus output from wind and PV plants, converted back to electricity when needed in a fuel cell or gas turbine. Similarly, spare renewable power can be used to compress air for later use in a turbine.

An entirely different balancing approach is to convert excess renewable output into heat and store that. It is much easier to store heat than to store electricity. However, it is hard to convert low-grade heat back to electricity. That is where combined heat and power (CHP) becomes useful. CHP plants can use either fossil gas or biomass/biogas as a fuel to generate electric power, and they recycle some of the waste heat produced in that process, so they have high overall energy-conversion efficiency. Crucially, the ratio of the heat to power outputs can vary depending on demand, so as to help with balancing. If there is too much green power on the grid, the CHP plant can produce more heat, less power. If heat demand is also low, the heat can be stored. If there is not enough green power on the grid, the CHP plant can increase its electric power output and reduce its heat output, and if demand for heat is also high, the stored heat can be used. Solar heat can also be fed into the stores and so can heat produced from surplus wind and PV electricity and from geothermal sources, all of this heat eventually feeding into local district heating networks.

As can be seen, the problem of renewable variability can, in fact, be turned into at least a partial solution. With a large renewable capacity on the grid, sized so as to meet most demand most of the time, there will at times be significant surplus electricity output, which can be stored and used, directly or indirectly, via the systems described above to meet lulls in renewable availability at other times. Electricity top-ups can also sometimes be obtained from overseas, using long-distance 'supergrid' links, so that local shortfalls can be balanced, making use of the likelihood that, in widely separated areas (with different weather conditions and time zones), demand and supply patterns will often vary out of phase. For example, local wind surpluses at one time can be traded with power from wind surpluses elsewhere at other times, as wind fronts move across the globe. The location and timing of peak solar input also shifts around the world as the earth rotates. Supergrids would also give access to large hydro pumped-storage reservoirs for those without them and possibly also to renewable sources far off, including large solar arrays in desert areas (Fraunhofer 2016).

This new supply and storage system would be complemented with a new demand-side management system, able to shift demand peaks to times when more power was available, for example by variable pricing so that power costs more at peak times. Optimizing it all will be hard, but as balancing technology develops it should be possible to move towards a balanced sustainable energy system at reasonable cost (see Box 2.1).

Box 2.1 Balancing costs

The technical costs of balancing variable renewables have been extensively studied, and one widely accepted estimate in the UK context is that they might add 10–15% to power generation costs at medium levels of renewable capacity, depending on what balancing technology is used, although it would rise at higher levels (Heptonstall, Gross and Steiner 2017).

At present, most balancing is achieved by using gas-fired plants, their output being ramped up and down to compensate for the variations in supply and demand, sometimes coupled with pumped hydro storage, if available, and batteries for short-term storage. That may be fine, with some extensions, up to maybe a 40–50% renewable power contribution: by 2019 the United Kingdom had reached 38%, Germany was nearing 50% and Denmark had reached 55%.

However, at higher levels, more may be needed. For that purpose, although they may well be the main elements, gas-fired backup plants (usually the cheapest option) and energy storage facilities (usually more expensive) are not the only balancing technology options. As noted in the main text, demand can also be managed via smart grid/variable energy pricing to delay energy demand peaks when variable renewable inputs are low, and top-up power can be imported to meet the peaks via long-distance supergrids.

Both those options can be low cost in operational terms. Indeed, flexible-demand management and smart grids can save money by reducing/shifting demand peaks, while supergrid links allow not just for balancing inputs but also for exports of surplus for some countries, earning a net positive income and avoiding the need for (wasteful) curtailment (or dumping) of surpluses. For example, a study by the UK government's National Infrastructure Commission claimed that an integrated flexible supply-and-demand management system, with smart grids, storage and also grid interconnector imports/exports, could save the United Kingdom £8 billion per annum by 2030 (NIC 2017). A study by Imperial College/OVO Energy claimed that just adding residential flexibility in domestic energy use (including for electric vehicle charging) could reduce whole-system costs by up to £6.9 billion per annum or 21% of total electricity-system costs. It was suggested that these savings could more than offset the cost of upgrading the power system. That does seem credible for some of the options. For example, introducing variable time-of-use energy tariff charges requires no capital outlay but would lead to reduced peak energy use and user costs and also lower system costs (Ovo/Imperial 2018). In all, it has been suggested that, if fully developed, system flexibility and integration could save the United Kingdom up to £40 billion by 2050 (Bairstow 2019a).

As noted above, as the renewable proportion goes up, so do the balancing costs, dramatically so in some modelling, for contributions of 70%, 80% and above. So savings like this would be welcome. However, there could be more savings to come if renewables expand even *further*. While balancing costs will rise until most power demand is met from renewables most of the time, after that any further expansion of renewable capacity, while requiring capital investment, will not incur extra power grid-balancing/backup costs. It will actually *reduce* the need for backup (more power would be available more often), while increasing the surplus that will be generated at times of low demand. The extra surplus would not be needed for balancing but would be available for heating, transport or export, or maybe conversion to hydrogen for these purposes, if that was the most lucrative option. In the latter case, more power-to-hydrogen conversion plants would be needed, but in either case the costs would be offset by the earnings from these end uses and the reduced system-balancing costs.

So a low-energy cost/high-renewables future may be possible, even given the need for balancing. That is certainly what a range of new scenarios propose, even those extending to supplying all energy, not just electricity. For example, the updated 2050 scenario produced by Professor Mark Jacobson and his team at Stanford University in California suggests that a system supplying 100% of global energy from renewables will not cost more than the current system and could actually be cheaper per kWh, even given the use of variable sources. Moreover, since it would avoid the increasing cost of fossil fuels and also the social and environmental costs of using them, it could be significantly cheaper overall (Jacobson et al. 2017).

Similar conclusions have emerged from studies by LUT University in Finland in conjunction with the Energy Watch Group (EWG) in Germany. They claim that 100% of energy, globally by 2050 or even earlier, is possible and would not cost more but in fact slightly less in direct cost terms: energy-generation costs would fall from €54/MWh for the system used in 2015 to €53/MWh with the new system, with balancing/storage, in 2050 (Ram et al. 2019). Note that neither the United States nor the European group saw nuclear as playing a role, not least because, as well as being expensive, it is inflexible and unable to balance variable renewables.

Making cost predictions so far ahead is obviously hard, and there have been queries about the use of projected *average* global capital costs for the calculations (Egli, Steffen and Schmidt 2019), given that there may be important local variations. However, that is difficult to predict, whereas the LUT researchers believe global-trend projections may be more reliable (Bogdanov, Child and Breyer 2019).

How rapidly can all this happen?

Although these scenarios sound very positive, and the supply and balancing costs look manageable, can the expansion of renewables for all energy, not just electricity, really be achieved on the *timescale* they suggest? As noted in chapter 1, some critics think not. For example, leading energy analyst Vaclav Smil concluded that 'replacing the current global energy system relying overwhelmingly on fossil fuels by biofuels and by electricity generated intermittently from renewable sources will be necessarily a prolonged, multi-decadal process' (Smil 2016).

New technology development, and more so system change, takes time, but the view that it is inevitably a slow process has been challenged (Lovins et al. 2018; Sovacool 2016). It has been argued that the transition to 100% renewable electricity could occur much more rapidly than suggested by historical energy transitions (Diesendorf and Elliston 2018), with concerns about climate change helping to speed the process.

The recent pace of development and take-up of PV and batteries, as well as electric vehicles, certainly suggests that change can happen quickly. Although there has been no shortage of speculation over the likely impact of 'destructive innovation' of this sort on energy industry incumbents, there have been proposals for the *very* rapid expansion of renewables in response to what some have portrayed as a climate emergency, for example to around 80% of UK electricity by 2030 (Greenpeace 2019). The global Extinction Rebellion campaign even called for 'zero carbon' by 2025 in an attempt to shift the definition of what is politically possible, so as to make it more in line with what is deemed scientifically necessary for ecosystem survival (ER 2019).

However, although the public mood may be changing, especially amongst the young, and renewable growth continues, it is wise to be a little cautious about what can be done in practice and how quickly it can be done: it may take time, and the political context sometimes does not support too much optimism. Support levels for renewables have been cut in many countries so, some say, the initial subsidy-based boom may falter.

Certainly, although investment levels have risen over the years, they have recently fallen off (BNEF 2019; IEA 2019b). Part of that may be due to the fact that new projects are cheaper, so less capital is needed to get the same energy output and economic return. Nevertheless, with less investment, the overall rate of capacity growth has slowed, with annual additions falling. Even so, *cumulative* capacity and output are still rising, and that looks set to continue in the years ahead (Wartsila 2019). Moreover, given the political will, it could be accelerated. For example, a study by the German BDI industrial forum suggested that getting renewables to near 90% of power by 2050 in Germany as part of an 80% greenhouse gas emissions-cut scenario could be technically feasible with the necessary support. Going further to a 95% emissions cut, with renewables supplying 100% of electricity, and also meeting other energy demand, was conceivable but was likely to be very expensive. The BDI said that would be challenging socially and also economically if Germany tried to do it alone without other countries adopting similar approaches (BDI 2018).

I will be looking at the overall cost of transitions like this in chapter 8 but, for Germany, the BDI put the overall net additional investment needed, set against the likely savings, at around €470 billion and €960 billion respectively (for the 80% and 95% emissions cuts) by 2050, or roughly €15 billion and €30 billion per year, around 0.4–0.8% of Germany's gross domestic product (GDP).

That is all a little speculative and some way off, whereas for the moment the reality is that, although funding programmes are continuing, investment-level growth in Germany and elsewhere is falling. Some critics argue that the recent fall in investment is due to the realization that renewables are expensive and that supporting rapid expansion with subsidies passes on unsustainable costs to consumers or taxpayers. That argument has been used in Germany, where guaranteed-price feed-in tariffs, which were very successful at building up renewable capacity, have been cut back and replaced by competitive project auctions in order, ostensibly at least, to cut the subsidy cost to consumers. The result has been a slowdown in renewable capacity build, which some see as the real aim, given the apparent hostility of some of the energy utilities to rapid renewables expansion that was undermining their markets. The economics of renewables can certainly be contentious: for example, they are allegedly getting cheaper, so why have consumer costs risen (see Box 2.2)?

Box 2.2 System costs, balancing and renewable economics

According to an OECD/NEA report, between 2008 and 2015 the use of low marginal-cost variable renewable energy had 'caused an electricity market price reduction of 24% in Germany' (OECD 2018). However, while wholesale costs may have fallen, this has not always resulted in the savings being passed on to consumers in terms of reduced retail prices, with residential electricity tariffs having increased in Germany by 16% between 2010 and 2017. It has been similar in the United Kingdom and in some cases elsewhere.

Power costs are becoming increasingly controversial, but it is also complicated to assess the factors leading to rises. The power utilities argue that the retail cost hikes have been because they face increased overall system costs, some of this, it is sometimes claimed, being due to the costs of the various green levies. They certainly have added some extra cost to bills, although that has to be set against the energy and cost *savings* from some of the green energy schemes. For example, the UK government's advisory Committee on Climate Change has claimed that the costs of green power subsidies to consumers have been more than offset by savings they enjoyed due to the various energy-efficiency initiatives (CCC 2019). I will be coming back to that issue in chapter 5.

The power utilities also complain about system-balancing costs incurred in managing the system with increasing amounts of variable renewables on the grid. Conventional plants may have to operate less efficiently to cope with variable renewables on the grid, ramping output up and down regularly. As has happened in Germany, some conventional plants will be forced to go offline for a while, so losing income when lower marginal cost renewables are available, the market for gas plants thus being undermined by smaller-scale, often locally owned, renewable projects. These extra costs/losses have sometimes been called 'market profile' costs since they result from the changed pattern of generation in the new power market, adding to the overall system-management and subsidy costs.

However, these various system costs will not have been the only extra costs faced by the power companies. Fossil fuel costs have risen and so have operational costs generally. Set against that, while the big utilities may be losing some of their markets to green energy in some countries, there are also savings. For example, about €8.8 billion of primary fuel import costs in Germany were avoided in 2015 as a result of renewable energy use, and these savings will grow as renewables expand, while the system-balancing costs should begin to fall as new smart systems are adopted, replacing the use of fossil plants for backup (see Box 2.1).

Meantime, gas turbine plants can offer a relatively cheap balancing option. In the United Kingdom, the government has introduced a capacity market system to ensure that sufficient balancing capacity is available, offering a subsidy to support it. However, while other balancing options (including storage) were also eligible for support, so far this system has focused mainly on gas plants and also, somewhat perversely, on existing nuclear capacity, despite this not being very useful for balancing variable renewables.

You might see the subsidy payments under the UK capacity market as being interim compensation for the 'market profile' losses of these existing plants. While it is hard to justify the inclusion of already subsidized and inflexible nuclear in this, it is important to keep flexible gas turbines online to help with balancing, although gradually other balancing options, including storage, interconnectors and demand management, will hopefully get more support. That may also happen in Germany, under market pressures, with utilities focusing on what could become a lucrative balancing market rather than on supply, much of which (although, as indicated later, perhaps not all) will come from local self-generation by 'prosumers' and energy co-op projects.

Clearly, whichever side of the debate you are on, the subsidy issue can be provocative. Some say subsidies are vital to help new technologies to enter well-established markets; others see them as undermining competitiveness and leading to extra costs. For example, a Chicago University study claims that the US Renewable Portfolio Standards (RPS) 'significantly increase average retail electricity prices, with prices increasing by 11% (1.3 cents per kWh) seven years after the policy's passage into law and 17% (2 cents per kWh) twelve years afterward' (EPIC 2019).

The RPS system is based on targets and market-determined prices, and, as the Chicago University paper says, the latter reflect 'the costs that renewables impose on the generation system, including those associated with their intermittency, higher transmission costs, and any stranded asset costs assigned to rate payers'. The latter outcome, where some old plants sometimes become uneconomic and perhaps have to be subsidized by consumers to keep them going, reflects the fact that renewable *generation* costs are low and getting lower, so that, as in Germany and the United Kingdom, some conventional plants cannot compete in the changed market (see Box 2.2).

Is it fair to include that changed so-called 'market profile' cost as a cost of renewables? It is arguably just a commercial loss or cost faced by their rivals. True, grid-balancing costs are an extra but arguably they reflect the cost of moving away from an inefficient, inflexible, centralized system to one based on distributed variable green sources and flexible backup and demand management. It is claimed by supporters of the latter that, although it may take time to bed these new systems in, if it is done right the new system should lead to more efficient supply and demand balancing and overall cost savings (see Box 2.1). Given also the emissions savings that the new system can deliver, the optimists look ahead to an unstoppable change process, with green energy technologies replacing the old technologies in an increasingly productive and competitive way (Lovins and Nanavatty 2019).

Some key strategic issues

Clearly that process will take time and may be disruptive, with some collateral costs and problems on the way. However, the operational and integration problems, and system costs like those looked at above, are to some extent the result of the *success* of renewables: they are displacing other sources, and it will take time to adjust the system.

There are also other problems of success. For example, as renewable costs fall, it may become less attractive to invest in new capacity since profit margins may be squeezed. Indeed, in competitive market systems, there can be a 'race to the bottom' to the point when there is little economic incentive to continue with new projects (Holder 2018).

This so-called 'market cannibalization' effect can be an issue for any successful product, but it may not be a 'showstopper' if demand for the product is expanding, as is the case with green energy. However, the competitive 'race to the bottom' can still lead to problems with speculative price bids. For example, winning bids for contracts for renewables in competitive capacity auctions may sometimes be set at low power prices that the project cannot in the event deliver, so that it does not go ahead. That happened with some early renewable energy projects under the United Kingdom's Non-Fossil Fuel Obligation competitive price/capacity auction system in the 1990s, and it may be a risk under the competitive auction systems now being used for renewables around the world (see Box 2.3).

Box 2.3 Market cannibalization and the race to the bottom

There have been some amazingly low-cost solar PV projects winning competitive power purchase contracts in capacity auctions around the world. For example, in 2017, solar auctions in Mexico yielded an unheard-of average price of $20.57/MWh, including a $17.7 bid by Enel, this beating an earlier $17.9/MWh tender for a 300 MW PV plant in Saudi Arabia. The price falls have continued. For example, in 2019, PV projects were given contracts in Portugal at €14.8/MWh ($16.6).

Low prices like these, comparable with, or sometimes even lower than, conventional power prices, may be exceptional and locationally specific. However, there has clearly been a trend to ever-lower bid prices, some of them perhaps being set unrealistically low, in some cases speculating against future possible energy price trends (Davis 2017a; Thurston 2017). Rules about project delivery, with fines for non-compliance, may need to be toughened up to avoid undue speculation and the risk of project failure. So there can be a downside to relentless competition. Interestingly, in this context, Mexico has recently halted its private contract auction process, and is looking instead to more of a role being played by the state power company.

All that said, there is no doubt that technology costs are falling, for PV especially, around the world. Lazard's 2018 review put the US levelized (lifetime-averaged) cost of energy for utility-scale crystalline PV, without subsidy, at $44/MWh. It noted that subsidies reduced the cost of power supplied to $32/MWh (Lazard 2018b). Even in the United Kingdom, not known for its sunny weather, contracts for large-scale PV (under the CfD system) were offered at £50/MWh ($65/MWh) and there has been talk of unsubsidized projects getting down to £40/MWh by 2030 (Stoker 2018). The trend seems likely to continue. Bloomberg New Energy Finance have suggested that PV costs will fall by 71% by 2050 (BNEF 2018a) and some even look to prices in effect bottoming out so that, at least for domestic supplies, the costs will be negligible (Clark 2018).

That may be very optimistic, but it is possible that, as has happened with some consumer ICT electronics, costs will fall to the point where the market has focused more on charging for the software and content than for the hardware. In the case of PV, that might mean that the emphasis for supply companies shifts to charging local decentral 'prosumers' for energy management services, to optimize supply-and-demand grid balancing and to aid peer-to-peer trading of surpluses. In time, a new type of energy service market may thus emerge.

While market-related issues will remain paramount, some of the issues facing renewables are more strategic, concerning uncertainties about the overall direction that should be taken. For example, in terms of scale, some look to large systems, seeking economies of scale, others to smaller projects, more appropriate to local communities. The classic large technology is hydro, the classic small technology is PV solar, although both can also be deployed at other scales, for example with large solar farms and small hydro projects.

As noted earlier, there are environmental issues with large hydro and tidal-barrage projects, but there are also economic issues with some small projects. For example, micro-wind domestic-scale units are very much less cost effective than large turbines, given that the power available is proportional to the *square* of the turbine radius, and large turbines can be located in wind farms in windier sites, with the power output being proportional to the *cube* of the wind speed. Solar PV is more scale-flexible, and domestic-scale projects have the advantage of delivering power direct to users without grid-supply losses or costs.

However, there are some economies of financing, deployment and operational scale, including for PV solar. For example, the cost/kW of large PV projects is lower than for small projects, with power utilities able to get access to low-cost investment capital more easily than homeowners. They can also build large numbers of projects, reducing the unit cost. The result is that larger utility-scale solar farm projects are usually much more competitive than domestic-scale projects (Brattle 2015; Lazard 2018b).

The often large economic-viability difference between domestic- and utility-scale projects may reduce somewhat as cheaper mass-produced PV cells and lower-cost storage technology emerge, and certainly, aided initially by feed-in tariffs in Europe and to a lesser extent net metering schemes in the United States, residential PV projects have flourished. Indeed, the power utilities sometimes complain that domestic PV users have enjoyed 'free rider' advantages since they do not pay the full grid-system usage and upgrade costs, although some PV advocates counter that the utilities are trying to squeeze domestic PV out of the market, for example by seeking to impose system-use charges. There has been a lively debate on this in the United States, in part focused on the level of payment for surplus PV power exported to the grid by prosumers. The same issue has also emerged recently in the United Kingdom, for example in relation to system charges and export tariffs for surpluses from domestic PV (Elliott 2019b).

While some look to renewables being developed on a smaller-scale local basis, and that is well underway in Germany and elsewhere, there are also wider operational and system-level constraints that may limit how much can be done at that level. Most obviously, not everywhere will have access to sufficient renewable inputs to be locally self-sufficient, so they will have to import some of their power. That may be necessary in any case, in order to balance local variations in demand and supply. Although local storage can help, trading power between different locations may be necessary and may also involve buying in power from some quite large projects, of the sort that can only be developed in specific locations, such as hydro, geothermal, tidal and wave projects. Nevertheless, there will be roles for smaller local projects: a balance will need to be struck (see Box 2.4).

Box 2.4 Scale issues: is small beautiful?

Local small-scale generation can deliver power direct to users, avoiding the need for the long-distance transmission and associated energy losses. It can also, in some situations, allow users to have direct control and ownership of the power system. Certainly, consumer-led PV uptake and self-generation by 'prosumers' and local energy co-ops, as has now spread across Germany and in some other countries, represents an archetypical 'destructive innovation', challenging and changing market and industrial paradigms. Schleicher-Tappeser, writing in the journal *Energy Policy*, said that it allows consumers of all sizes to produce power themselves: 'new actors in the power market can begin operating with a new bottom-up control logic.' He added that the 'increasing autonomy and flexibility of consumers challenges the top-down control logic of traditional power supply and pushes for a more decentralised and multi-layered system' (Schleicher-Tappeser 2012).

That harks back to the radical agenda outlined by the late Hermann Scheer in his 2005 *A Solar Manifesto*: 'Since everybody can actively take part, even on an individual basis, a solar strategy is "open" in terms of public involvement. It will become possible to undermine the traditional energy system with highly efficient small-technology systems, and to launch a rebellion with thousands of individual steps that will evolve into a revolution of millions of individual steps' (Scheer 2005).

Along these lines, a Greenpeace scenario suggested that in theory, in most places, up to 70% of energy could be generated and used on a local basis, with only perhaps 30% involving larger-scale systems and grid trading (Greenpeace 2015). However, this may be optimistic. The 70/30 ratio is rather idealized and seems unlikely to be viable everywhere. Renewable sources are not available to all to the same extent, and, at any specific location, there will be variations in availability over time. Greenpeace admits that a fully decentralized system would need more (oversized) local capacity in order to maintain stable supplies than if you could rely on grid imports for balancing, so there could be a cost implication.

Local generation, aided by onsite storage, may nevertheless play a key role. It can reduce the strain on the grid and the need for new centralized generation capacity. There would be less demand on the grid during times when consumers/local communities could use their own power, although at other times they would still need power, leading to a capacity-balancing requirement.

The need for backup at times is of course why going *fully* off-grid may be problematic. That can be done, and is done in some remote locations, with the advent of cheaper batteries making it easier. For some consumers, even in areas with grids, going off-grid may still look appealing since it holds out the promise of individual 'self-sufficiency' and being free of reliance on the power utility companies. However, at present, trying to go for *total* energy autonomy at the household level may make little sense technically or economically in locations where there are grids for backup and, crucially, for trading any surpluses. Small isolated private power systems are likely to cost users more and are also arguably less environmentally sustainable and efficient than well-managed grid-linked and balanced systems. Nevertheless, if power grid systems begin to fail or become more expensive to use, for whatever reason, then more people may opt out.

While decentralized renewable energy supply, with self-generation of power by prosumers, including storage and peer-to-peer trading of surpluses, looks likely to become increasingly important, as argued above, there will also be a need for wider trading and the use of larger sources. The optimal balance between these elements is debated and will vary by location. However, the emphasis is likely to change from that at present. In its visionary UK projections, the UK Institute for Public Policy Research looked to a new smart energy system which 'empowers citizens and communities to be more self-sufficient while being part of a connected, inter-dependent system that offers security of supply and resilience in the face of changing demand and climate' (IPPR 2018).

In addition to wider strategic debates on issues like this concerning the mix and scale of renewables, there are some general issues in relation to other possible pathways ahead, for example the role of carbon capture and storage and its potential use with biomass. Biomass use has its supporters but, as already noted, also its opponents, who are concerned about the likely environmental impacts and land-use implications of growing energy crops on a large scale. The stakes have been raised since some supporters want to go for biomass combustion combined with carbon capture and storage (BECCS), so as to achieve a net *negative carbon* outcome, assuming the biomass used is replaced with new plantations and the CO_2 is stored.

Not everyone is convinced that biomass use can be fully carbon neutral, much less carbon negative, given that it takes time for new biomass to grow to re-absorb CO_2 and replace the lost carbon sinks, a key issue in terms of the use of forestry products, as I explore in chapter 3. It is also not clear if all the carbon emissions can be captured and stored effectively by carbon capture and storage systems on a permanent and wide-scale basis. Nevertheless, some still see carbon-negative options as vital. I will be looking at the carbon sequestration issues further in chapter 4.

Finally, there is the overreaching issue of whether we should be seeking to develop new energy supply and carbon reducing/storing technologies, of whatever sort, as opposed to technologies for avoiding the need for more energy supply. The more efficient use of energy can, arguably, avoid emissions/kW of finally used energy at lower cost than adding more supply. Certainly, energy saving and demand reduction ought to have been given much higher priority than they have so far. In most industrial countries, energy has in the past been *relatively* cheap and the social and environmental impacts of using it have been externalized (i.e. left to society to deal with). That situation has now changed, so there should be more of an incentive to avoid energy waste.

There are clearly significant potentials for energy (and carbon) saving. For example, a UK study claimed that 'one quarter of the energy currently used in UK households could be cost effectively saved by 2035; and this could increase to one half if allowance is made for falling technology costs and the wider benefits of energy efficiency improvements' (Rosenow et al. 2018).

Similar gains are also possible in all the other sectors, including industry and transport. Overall, a recent report from the Centre for Research into Energy Demand Solutions at Oxford University sees improved energy use efficiency as being the key to the decoupling of energy demand from economic activity, thereby cutting emissions, much more so than clean energy supply. 'In recent decades, more than 90% of the progress in breaking the relationship between carbon emissions and economic growth globally has come from reducing the energy intensity of the economy. By comparison, reducing the carbon emissions per unit of energy has, to date, been a relatively minor effect' (Eyre and Killip 2019).

However, that may overstate the role of efficiency. Some of the reductions in energy use have been due to structural changes in the economy and to the rising cost of energy, not to improved energy efficiency per se, issues I will be returning to in later chapters. There may also be conflicts between energy saving and energy supply. The increasingly low cost of renewables may undermine the economic attraction of energy saving. In some situations, it may be cheaper (per tonne of carbon avoided) to invest in green power supply than to invest in energy efficiency, especially once all the easy, low-cost energy savings have been achieved. That is debated: some say there will be economies of market scale as energy-saving techniques develop and are widely adopted (Lovins 2018).

Also much debated is the potential impact of the so-called 'rebound effect': the money saved by investing in energy efficiency may be re-spent on other uses of energy, so wiping out some of the energy and carbon savings (Wei and Liu 2017). Unless, that is, the money is spent on renewable power. Then the carbon savings from efficiency will be fully captured, a point I made a while ago (Elliott 2004).

Nevertheless, issues like that aside, it still makes overwhelming sense to avoid energy waste, and there are also valuable synergies between renewable supply and energy saving: if combined, they can limit the rebound effect. In addition, if demand can be reduced and also managed flexibly, it is easier to meet it from renewables. Moreover, since there are some impacts from using renewables and the renewable supply technologies also have material requirements, it is foolish, in resource and impact terms, to waste the energy that they can supply and then have to generate more.

The bottom line is that we need to pay attention to both the supply side and the demand side. It will be hard for renewables to meet demand unless that is reduced but, equally, even if demand is reduced dramatically (Germany is aiming for a 50% cut by 2050), we will still need carbon-free supply. Some say that the balance should tip towards demand-side initiatives, and certainly in the past energy suppliers have not seen it as commercially sensible to support energy saving. In the new environmentally constrained context, the balance definitely needs to be changed. However, that should not be too difficult. In the new context, in most cases, with possible exceptions as noted above, the supply and demand sides are not in conflict: both are needed for a sustainable energy future.

Choices ahead

In the next chapter, I will look at how these various strategic issues have been dealt with in some of the long-term scenarios and plans that have emerged. Some analysts stress the demand side and, as noted above, that is clearly very important. For example, the Oxford Environmental Change Institute's director, Professor Nick Eyre, has said that 'The goals of a secure, affordable, low-carbon energy system are only achievable if energy demand is reduced, decarbonized and made more flexible' (ECI 2018). But even in that formulation, decarbonization is presented as a key element, and that means new low- or zero-carbon supply technologies.

Nuclear power is one such option, but not one that I have covered here in much detail since, as I have argued at length elsewhere (Elliott 2010, 2017b), it has many drawbacks, including economic, safety and security issues and the problem of long-lived radioactive waste. There is also the already mentioned more practical problem that nuclear plants are inflexible and will not be much use for balancing variable renewables. They just get in the way of the more flexible supply and demand system that will be needed for a renewables based system. There have been proposals for smaller, more flexible nuclear plants, but that is some way off, their economics, as well as their safety and security risks, still being uncertain (Thomas et al. 2019).

While some still look to nuclear as an interim option, possibly using new technology, in the longer term there is the fundamental issue that nuclear fission relies on fuels that, like fossil fuels, are a fixed planetary resource. They are not being renewed and cannot be relied on indefinitely. Indeed, although estimates vary, current reserves seem likely to be sufficient to supply the global reactor fleet for only a few decades, optimistically maybe up to a hundred years or so, although less if, as some would like, the use of nuclear is expanded. New uranium finds may be made as costs rise, and new technologies, like breeder reactors and the use of thorium, can extend the use of the fissile resource, but nuclear fission is not a *renewable* option.

The fuel sources for nuclear fusion, if it is ever success-fully developed at commercial scale, are more extensive, with one key hydrogen isotope fuel, deuterium, being available from seawater, although for the other, tritium, there may be limits to its availability on this planet. There are other uses for lithium, the main current source of tritium, including for the batteries of electric vehicles. Recourse might be made, in time, to helium-3 or other reserves off-planet, and that might be necessary if we were to try to use fusion for powering space vehicles, which otherwise will have to continue to use mainly fossil-derived fuels. However, that is all some way off, and major breakthroughs and science fiction-type futures aside, quite apart from having safety and materials irradiation issues, fusion does not seem to offer a solution to the urgent need for non-fossil fuels (Elliott 2019c).

Looking far ahead is fraught with difficulties, but the prospects for renewables do look much brighter and they are available now, unlike fusion, offering us a way to respond quickly to the climate change threat. Specific technology choices for supply and end-use management will be hard to make, but some sort of general consensus seems to have emerged in the wider picture. It does seem likely that, if we want to go that way, renewables can supply the bulk of global power by around 2050, and also heat and transport fuel, if proper attention is paid to energy saving and demand management. Nevertheless, there are differences in detail and emphasis; there is as yet no one fixed optimal energy future, but rather a series of potential routes forward, with a range of possible technologies being used. In the next chapter, I look at which options have been chosen so far and which have been or might be sidelined.

3
Energy Technologies for the Future

Having set out the technical options and issues in the previous chapter, in this chapter I look at which options have been chosen to aid the transition to a sustainable future in some key future scenarios and also sets out some of the options that are less popular.

Multiple possible futures

Energy scenarios map out possible mixes of technology and how they might meet expected demand. There are many such scenarios and studies, some, as I have indicated, looking to very high renewables contributions. In fact, according to one survey, there are 42 peer-reviewed academic studies so far from around the world showing renewables supplying up to 100% of all electricity nationally or globally by around 2050, or even, in some cases, 100% of *all energy* globally by 2050 (Stanford 2019).

However, not all energy analysts see renewables as being able to achieve levels like this. As noted earlier, some see fossil fuel and to a lesser extent nuclear still playing major roles. For example, a recent World Energy Council (WEC) study reviewed scenarios from Shell, Statoil, Exxon, IEA, IPCC, IRENA and others, alongside its own scenarios. It found that the trend projection-based scenarios for 2040 all showed renewables at less than the 30–40% of the energy mix achieved in the highest 'normative' new policy-driven scenarios, in some cases a lot less. Moreover, nearly all scenarios depicted energy demand rising, with fossil fuels still supplying 70% or more of energy by 2040, so that oil, gas and in some cases coal stayed in the game, as did nuclear to some extent (WEC 2019).

There are limits to the reliability of scenario modelling. Those based on simple extrapolation of trends can be confounded by technological breakthroughs or unexpected events, while those based on normative 'pathways' may prove to be technically, economically or politically unviable. In most cases, attempts are made to base the projections on the best data available and to carry out careful assessments of feasibility but, perhaps inevitably, quantification and computer modelling can lend a spurious certainty to the results.

All that said, it is helpful to quantify and test possibilities as far as possible, so as to ensure that, as the late Professor David MacKay said, 'the sums add up', something he tried to achieve in his influential text (MacKay 2013). While that is important as an aim, in practice, at this stage, given the open-ended nature of the emerging energy system, it is often hard to come up with reliable data to feed into the modelling. For example, some of MacKay's data now looks dated (Goodall 2017). Furthermore, modelling itself, and its framing, can have problems. It will inevitably reflect the underlying implicit or explicit assumptions made.

Bearing that in mind, it might be argued that the scenarios which WEC looked at were rather conservative, in which case it is interesting to also look at the more optimistic longer-term scenarios and plans, which the WEC review did not cover, such as those produced by academics and NGOs. In those, renewables expand more rapidly, wind and solar especially. For the other green energy options within these scenarios, there are differences in emphasis, some of them reflecting the strategic issues discussed in the last chapter. Thus, in some cases, biomass use is mostly avoided, while in others hydro is downplayed. Some see energy saving playing more of a role; in others, battery storage is a major element, while others look to interconnectors for balancing, along with demand-side management. Negative carbon options are sometimes seen as part of the mix, although few of the mixes include nuclear. Even so, there is no one fixed future, just a range of scenarios, each with different implications. This chapter takes a look at some of the global 'high renewables' scenarios and plans, focusing on the leading ones from Jacobson's team in the United States and from the LUT/EWG in Europe, and highlighting the mix of technologies that has been selected in them.

Wind and solar PV dominate

By 2050, wind and solar PV will account for 96% of the total power supply of renewable sources, according to the LUT/EWG '100% renewables by 2050' global scenario, mentioned in chapter 2 (Ram et al. 2019). So, on this view, they will dominate all other renewables and also supply the bulk of global electricity.

Wind is currently leading in capacity terms, heading for 1 TW globally, but PV seems to be getting cheaper faster. However, while there are strong advocates of both, it is not really a contest: it is widely accepted that we need both, not least since they complement each other well. Wind has the advantage of being available at night and is at its strongest in the winter, but PV is easier to install in more locations, including near users, with low maintenance requirements and no moving parts to service. It is also well suited to meeting daytime air-conditioning loads in summer, and with climate change demand for that will grow.

The costs of both wind and PV are still falling, and rapid continued growth does look possible, with the assumed scale of wind and (especially) PV solar expansion in some scenarios being quite staggering. In the initial global scenario from Jacobson's team at Stanford University, 13 TW of wind capacity will be in place by 2050, with onshore wind supplying 23% of total final end-use power and offshore wind 13.6% (Jacobson et al. 2017), expanded in Jacobson's new scenario to 15.6% (Jacobson 2019a). In the longer term, for example if airborne devices are successful, it could be very much more. Nevertheless, the potential for PV is usually assumed to be very much greater than for wind. In the 139-country 'all energy' 100% renewables scenario from Jacobson and colleagues, solar is at around 46%, with nearly 30 TW of PV in place by 2050 (Jacobson et al. 2017). Jacobson has provided a helpful graphic guide to the scenario illustrating the way it can all work together (Jacobson 2019b).

In scenarios from LUT/EWG, PV plays even more of a role (and wind less of one). Their all-energy sectors global scenario predicts an astounding 63 TW of PV by 2050. In all, in this scenario, solar PV generates 69% of global primary energy, followed by wind energy at 18%, biomass and waste at 6%, hydro at 3% and geothermal energy at 2%, all by 2050 (Ram et al. 2019).

So, on this view, solar clearly leads, and in terms of avoiding land-use conflicts, interestingly, LUT researchers have estimated that globally there could be 4.4 TW of floating PV arrays on just 25% of the surface area of hydro reservoirs and more on other reservoirs (Farfan and Breyer 2018). That said, some very high estimates for wind potentials have also emerged recently, with a GSI-based study claiming that, taking account of socio-technical constraints, there was room, in theory, for 52.5 TW of onshore wind in Europe, which, if its output could be distributed, would be sufficient to cover *global* all-sector energy demand by 2050 (Enevoldsen et al. 2019).

Some of these projections may be unrealistic, especially given local constraints (e.g. on the very widespread deployment of wind or solar farms), and some estimates from industrial consultants are much less optimistic (Anderson 2019; D. Brown 2018). Even so, it seems clear that wind, and solar especially, will prosper, as has been recognized even in some of the scenarios from oil companies that retain significant amounts of fossil fuel use and, to a lesser extent, nuclear input. For example, in its longer-term 'Oceans' scenario, Shell describes a pathway in which solar grows to become the largest single primary energy source in the energy system by 2060, accounting for up to 30–40% of total primary energy. Its latest very ambitious 'Sky' scenario has renewables, including hydro and biomass, then at 67% of the energy mix, led by solar in 2070.

However, while they all agree that nuclear will not expand much, not all oil company scenarios are positive about the role of renewables, nor the prospects for wind and solar, especially in the shorter term. Instead, they see fossil fuels, gas especially, continuing to play major roles (see Box 3.1).

Box 3.1 Some other oil company views on renewables

In its 2018 '2040 Outlook', BP sees renewables as supplying only around 14% of global energy by 2040, led by wind and solar, or about 20% with hydro included, although it also explores a 'renewables push' scenario in which extra support results in renewables supplying around 85% of *electricity* by 2040, up from about 45% without the push (BP 2018b).

Exxon Mobil is much more conservative. It sees renewables as marginal in *energy* terms, so that wind, solar and biofuels reach only about 5% of global energy demand by 2040, and even adding the 3% or so it estimates for hydro by then only puts the total at around 8%, although it says renewables would supply about 30% of global *electricity* by 2040. Like BP, it still sees fossil fuels supplying most non-electric energy. Although oil use for light vehicles might peak by 2030, it says that oil will continue to play a leading role in the world's energy mix, with growing demand driven by commercial transport and the chemical industry as a key feedstock (Exxon Mobil 2018).

Recent trends, including heightened concerns about climate change and about the impact of plastics on the environment, may be having an impact. But with energy demand seen as rising, BP's 2019 'Energy Outlook' still predicts renewable energy only making a 15% *primary* energy contribution by 2040, or 22% including hydro. Non-hydro new renewables do grow rapidly in its scenario, faster than anything else, and coal and oil take a big dive, but gas remains king. Nuclear stays low. But even in BP's *rapid transitions* scenario, renewables still only hit 29% (38% with hydro) of global energy by 2040 (BP 2019).

In chapter 5, I will be looking at some of the wider reasons why the proportion of renewables might be low in future: the most obvious one (as assumed in the oil company scenarios looked at in Box 3.1) is that energy demand may continue to grow, so renewables may not be able to increase their share much.

In terms of more *direct* reasons as to why their expansion may be constrained, one possibility is that some of the renewable options may face local opposition. There have been issues with wind farm and solar farm deployment in some locations but, in general, wind and solar have proved to be very popular with the public, and that support is increasing. For example, a 2017 public opinion poll in the United Kingdom showed 84% supported PV solar, 79% backed offshore wind and 74% supported onshore wind. In a 2018 UK survey, support had risen to 87% for PV, 83% for offshore wind and 76% for onshore wind. In 2019, support rose further to 79% for onshore wind and 89% for PV, but stayed at 83% for offshore wind, with overall support for renewables reaching 84%. Opposition to renewables has remained very low, at around 3%, over the whole period (BEIS 2019a).

The situation elsewhere seems to be similar. Amongst American adults, 83% supported expanding wind farms, while 89% supported solar expansion, according to a Pew Research Center study, although interestingly in some locations wind was preferred to solar (Firestone and Kirk 2019). Globally, of those asked in a global public opinion poll in 2017, 82% backed renewables, 80% wanted more solar, 67% more offshore wind and 64% more onshore wind (Orsted 2017).

Although attention does have to be paid to careful public consultation and siting, if that is done properly public opposition seems unlikely to slow wind and solar much. There will, however, be limits. For example, in land-use terms, PV solar is best deployed on the rooftops of houses, warehouses or factories, not on the ground, although some say that, even if farmland is used, some of it is still available for grazing and for wildlife. Debates over that continue, particularly in the densely populated United Kingdom, with some arguing that, apart from rooftops, only brownfield sites and marginal land should be used, and certainly not high-quality farmland (Elliott 2019b). Then again, the total area that might be needed for even quite large power contributions is relatively small: it has been claimed that 10 GW of PV installed on the ground would only use around 0.1% of UK agricultural land (STA 2019). By comparison, roughly 0.5% of UK land is currently used for golf courses. The land-use constraints in other, less densely populated countries will be much less.

With regard to wind, the issues are somewhat similar, although in this case the best, windiest sites are usually remote from human habitation, most obviously those for offshore wind. While sites can be contested, direct conflicts can usually be avoided and, for onshore wind, it is possible to farm right up to turbine bases, so reducing local land-use impacts.

By contrast, there may be more fundamental issues for some of the other renewable energy options, and there are divergences in opinion concerning which renewables to use and which to avoid, especially in relation to biomass and hydro.

No to biomass?

Most scenarios use a mix of all options and, in nearly all, wind and solar play major uncontested roles, but the very ambitious scenario produced by Jacobson and others, which as I have noted looks to obtain all of the world's energy from renewables by 2050, explicitly excludes most biomass use.

The scenario's lead authors say bluntly that:

> any use of land for the production of bioenergy feedstocks is worse for climate, water quality, soil, biodiversity, and overall ecosystem health than is the always-available option of restoring land to its ecologically best use and getting energy from other (non-biomass) sources. Put another way, getting energy from wind, water, or the sun rather than from bioenergy allows society to put land to better use than growing energy crops. (Delucchi and Jacobson 2016)

As I noted earlier, it is true that biomass use can have significant environmental impact and land-use issues. However, there can be problems with blanket opposition if applied inflexibly across the board. Biomass is a potentially very large global energy resource, a key attraction being that it can be stored, and although there are clearly issues with its use, some of them can hopefully be avoided or limited. For example, as already noted on p. 9, the use of farm wastes and domestic food wastes for biogas production via anaerobic digestion (AD) involves no new land use, and capturing the bioenergy content in these wastes avoids the emission of methane, a powerful greenhouse gas, which would be produced and released if they were just left to rot in the air or in uncapped landfill sites. In fact, some of these wastes are already widely collected and used for AD biogas generation, and capped landfill-gas sites, along with sewage gas plants, provide some of the cheapest sources of renewable power. A recent study by the Global Biogas Association says that only 2% of available feedstocks undergo anaerobic digestion and are turned into biogas and that, if developed, biogas could cut global greenhouse gas emissions by up to 13% (Bairstow 2019b).

In a recent update, Jacobson has in fact indicated that he accepts that some wastes could be used. He supports using 'landfill and digester methane to produce hydrogen by steam reforming, where the hydrogen is subsequently used in a fuel cell' (Jacobson 2019c). Fuel cells convert hydrogen to electricity with no emissions other than water vapour, so the CO_2 that would be produced by burning biogas directly is avoided, but steam reforming of methane would still produce CO_2. It is hard to avoid CO_2 production, however biomass is used. For that and other reasons, Jacobson says biomass use is mostly to be avoided in favour of his chosen mix of wind, water and solar technologies. For example, he claims that 'combusting forest and industry residue and other forms of biomass to provide electricity and heat results in higher CO_2 emissions and much more air pollution emissions than WWS technologies. Some forms of biomass also require much more land than do WWS technologies.'

A somewhat similar view has been presented by the US Natural Resource Defense Council in relation to the United Kingdom's use of forestry-derived biomass pellets imported from North America for power production (NRDC 2016). It is backed up by a study that claims that the United Kingdom could get to 'high renewables' penetration *without* using any biomass (Vivid Economics 2018). So it seems it might be possible to do without it. But is that necessary? Certainly, there can be problems with using forestry products for power production. In theory, replanting and growth can lead to a net carbon balance over time but, in the short term, vital carbon sinks may be lost and, unless they are replaced rapidly, there may be more CO_2 in the air than would be the case otherwise.

In addition to this problem, biomass has a lower calorific value than fossil fuel, so more of it has to be burnt in tonnage terms to get the same energy output, potentially generating more net CO_2, depending on the type of biomass used and the replanting rate. Indeed, it has been claimed in a Chatham House report by Duncan Brack that, 'overall, while some instances of biomass energy use may result in lower life-cycle emissions than fossil fuels, in most circumstances, comparing technologies of similar ages, the use of woody biomass for energy will release higher levels of emissions than coal and considerably higher levels than gas' (Brack 2017).

That assessment was rejected as extreme by, amongst others, the IEA Bioenergy group (IEA 2017), but provocatively there were allegations that some of the wood pellets that the United Kingdom was importing were from whole trees or stemwood, and that the CO_2 sink impacts would be significant (Montague 2018). As I have explored at length elsewhere (Elliott 2019a), there has been a sometimes bitter debate on this and related issues in the United Kingdom. One result has been that some earlier estimates of biomass potential have been reduced (see Box 3.2).

Box 3.2 Some UK biomass assessments

In 2014, the UK Tyndall Centre for Climate Change suggested that it might be possible to obtain 44% of UK energy from biomass by 2050, including from energy crops and the use of food and farm wastes, with no imports being required and without impacting food production. It looked to residues from agriculture, forestry and industry potentially meeting 6.5% of primary energy demand by 2050, while waste resources were found to potentially provide up to 15.4% and specifically grown biomass and energy crops up to 22% of demand (Welfle, Gilbert and Thornley 2014).

However, there were net greenhouse gas (GHG) issues, depending on the type of biomass used and the impact of harvesting on carbon sinks. A UK government study noted that 'the GHG intensity of electricity generated from North American roundwood and energy crops varies significantly, depending on the carbon stock of the land and the counter-factual', i.e. what else the biomass might be used for. It stated that 'Some scenarios can have very low (even negative) GHG intensities, if they result in increased carbon stored on the land. However, other scenarios can result in GHG intensities greater than electricity from fossil fuels, even after 100 years' (DECC 2014).

So it was possible that some biomass use could be counter-productive. However, while some were convinced that forest products were an example of a poor choice, others claimed that, in reality, the use of stemwood (arguably the worst offender) would be rare, given that there were more lucrative commercial uses for it. Against that it was suggested that rising demand for wood pellets might divert forest products away from other, perhaps better, uses of wood (in terms of storing carbon) and lead to more regular and extensive harvesting, destroying natural carbon stores more rapidly. Clearly, a complex set of issues emerged, with conflicting views (Evans 2015).

In the somewhat more cautious policy context that subse-quently emerged, the government's advisory Climate Change Committee suggested that the United Kingdom might reasonably get 5–15% of total primary energy from biomass, including wastes, by 2050 (CCC 2018a) – much less ambitious then, with land-use/carbon-sink issues presumably being seen as a constraint.

Interestingly though, public reactions are apparently not a major issue. The general public, at least in the United Kingdom, seems quite enthusiastic about bioenergy. In an opinion survey by YouGov for the Energy Technologies Institute in 2016, 80% of respondents supported an increase in bioenergy use in the United Kingdom, 74% supported producing bioenergy from biomass and 81% backed producing biomass from waste, comparable to levels of support seen for other renewables (ETI 2016). However, the ETI did note that more than a third of respondents were concerned about biomass competing with other land uses such as food production.

The UK debate on biomass has continued and broadened but has become somewhat polarized into strongly pro- and anti-biomass camps with, for example, the Renewable Energy Association pushing bioenergy quite hard (REA 2019) and some green groups opposing almost all aspects of it (Garson 2019).

In theory, the use of forestry wastes (rather than whole-tree stemwood) could avoid some of the problems, and controls have been put in place to block unsustainable felling, but in practice power demand may be such that stemwood, if not whole trees, may end up being used illicitly. Similarly, with the global market for biofuels for vehicles being even more lucrative, there can be major issues, with vast monoculture palm oil plantations for biodiesel production putting a squeeze on food production. The food versus fuel debate has been if anything even more bitter than that over the use of forestry biomass for power production. Certainly, quite apart from land-use and biodiversity issues, the energy content of some vehicle biofuels (bioethanol from corn especially) can be very low so that, given that energy is needed for harvesting, transport and processing, the net gain in carbon saving may be small or even negligible.

Supporters look to the adoption of non-food biomass with higher energy content as an answer to the food vs fuel issue, but the energy yields may be lower and production costs may be higher. Moreover, that does not really deal with the land-use and biodiversity issues, or the loss of carbon sinks through the regular harvesting of bioenergy crops.

In general, what is needed in environmental and climate terms in relation to the use of biomass for energy is a dynamic system in which replanting and growth rates match harvesting/ use rates so that the net carbon-sink level stays as high as possible. However, this is the exact opposite of the 'slash and burn' approach that is often the most commercially attractive short-term biomass/biofuel option. So there are potentially some fundamental conflicts, at least with most of the biomass sources and use patterns that have been adopted so far. Until those are resolved, the general picture is one in which biomass production and use is likely to be more constrained than was originally hoped. This despite the likelihood that demand for biofuels, including for aircraft, will grow.

The result is that there has been a race to find new, less problematic biomass crops, sources and use patterns, including faster-growing plants. For example, there is some enthusiasm for short-rotation willow coppicing and for fast-growing eucalyptus and elephant grass. In addition, there is more emphasis on the use of wastes and even recycled cooking oil (as a biofuel), while some look to novel sources like algae as a way forward, possibly grown on ponds or in seaweed farms near shore, or even to cactus or agave grown in desert areas.

Meanwhile, cutting across much of this debate, as I noted in chapter 2, there has been some enthusiasm for bioenergy with carbon capture and storage (BECCS), as a carbon-*negative* option. The carbon dioxide gas absorbed from the air by the growing biomass would be released when it was burnt for power production but would be captured and stored, so that the overall process would be carbon negative, process losses aside. However, as noted in Box 1.2, to have a significant impact on atmospheric carbon levels requires large areas of biomass plantation and also large carbon dioxide storage caverns/strata. That may significantly limit its potential.

Nevertheless, BECCS has been talked up strongly as being necessary and urgent since some fear that renewables like wind and solar might not be able to deliver sufficient carbon reductions fast enough. In parallel, there has been enthusiasm for direct air capture and storage (DACS) of carbon dioxide. I will be looking further at that in chapter 4. However, in the present context, suffice it to say that, while DACS would avoid biomass use, it would require energy to run, rather than, as with BECCS, generating energy and, like BECCS, it would also need large spaces for CO_2 storage. So there are some interesting and difficult trade-offs, as I have explored in detail elsewhere (Elliott 2019d).

The viability of BECCS (and DACS) obviously depends on the successful development of fossil carbon capture and storage (CCS), but that has been making very slow progress, in part because of its high cost and the problem of finding secure carbon storage space. Most of the development emphasis of late has moved on to carbon capture and utilization (CCU), for example using captured CO_2, along with green hydrogen from renewable sources, to make new fuels, e.g. for vehicles. That could also be done with biomass, i.e. bioenergy with carbon capture and utilization (BECCU), or with direct air capture, although then the process would no longer be carbon negative. Moreover, it would only make economic sense if the new fuels were more valuable than the hydrogen needed to make them, or the renewable energy needed to make the hydrogen. Otherwise, it would be more economic just to use the zero-carbon renewable power, or the hydrogen, directly.

I will be returning to the carbon economics issues in chapter 4: some say that, with rising energy demand, we may need negative emissions technologies (NETs), including BECCS, to balance the carbon budget. Then again, some say that just planting more trees and other biomass would be a cheaper and easier carbon-negative option. That, along with new ways and patterns of farming and soil management to protect and expand carbon sinks, could be a more effective way forward than BECCS (Hausfather 2018).

However, there are obviously limits, for example to how much reforestation is possible in practice, given that it is hard at present even to resist deforestation. Clearly, we need a new approach to land use and arguably also to agriculture, but that opens some large non-energy issues. For example, should we change dietary patterns and move away from land-intensive meat production? That might reduce pressure on forest areas (e.g. in the Amazon basin) but it might also open up more space for energy crops.

Human beings have used biomass since early times, and around three billion still use it (wood, charcoal or dung) for cooking and heating, often very inefficiently and with significant health risks from toxic fumes. As yet, the use of modern biomass for power, heating and transport still only accounts for a relatively small proportion of global energy supply. Around 5% of total energy is supplied from bio-sources, with biomass supplying 3% of transport fuel and only 2% of electricity. As I have indicated, it does have a large potential but, given the problems, its expansion may be limited, at least in some usages. Certainly, some of the 100% renewables scenarios do not see biomass/waste use expanding much, if at all. LUT/EWG has it at 6% of global primary energy by 2050; Jacobson has almost none being used by then.

No to hydro?

By contrast, hydro currently supplies around 16% global electricity and some hydro supporters see that expanding very significantly, perhaps doubling by 2050. However, there are major environmental and strategic issues with large hydro projects, which have led to the suggestion that 'the era of the awe-inspiring mega hydropower projects such as the Hoover Dam in the US and the Three Gorges in China should be coming to an end in favour of smaller projects' (Sovacool 2019; see Box 3.3).

Box 3.3 Hydro pros and cons

A study by the Science Policy Research Unit (SPRU) at the University of Sussex, United Kingdom, and the International School of Management, Germany, suggested that 'current calls for substantial, global investment in hydropower installed capacity and generation, including those from major institutions such as the International Energy Agency, IRENA, IPCC and World Bank, must be closely scrutinized' (Sovacool and Walter 2019).

The study compared the security, political governance, economic development and climate change performance of major hydropower states against oil-producing states and all other countries using 30 years of World Bank data. It found that, although net emissions were reduced, countries relying on hydropower have seen poverty, corruption and debt levels rise and their economies slow at significantly greater rates than nations which have used other energy resources over the last three decades. In addition, carbon reduction benefits were realized only over time, after an initial environmental impact from construction, while the financial benefits of major hydropower projects could take decades to emerge, with large budgetary overspends often associated with major hydro projects.

The report noted that the World Commission on Dams had estimated that about four million people were displaced annually by hydro construction or operation, while another study, looking at global energy accidents over 100 years, found hydroelectric dams caused 94% of reported fatalities and $9.7 billion in damages.

On environmental impacts, the report noted that hydro projects certainly can have negative impacts on habitats, water quality and environmental sustainability, but they also have positive impacts in terms of avoiding emissions. In that context, interestingly, the report cited the assertion that hydro reservoirs can:

> become virtual methane factories, with the rise and fall of the water level in the reservoir alternately flooding and submerging large areas

> of land around the shore; soft green vegetation quickly grows on the exposed mud, only to decompose under anaerobic conditions at the bottom of the reservoir when the water rises again. This converts atmospheric carbon dioxide into methane, with a much higher impact on global warming.
>
> But it said that, overall, net negative emissions impacts were not found in the study and that hydro 'reduced greenhouse gas emissions per capita'.
>
>   Nevertheless, it is still a conflicted technology, attracting much opposition, large projects especially, not least because of the local disruption they can impose. The report argued that smaller-scale, run-of-the-river designs that can operate without reservoirs, as deployed in Nepal, Tanzania and Sri Lanka, could be used more widely to limit environmental problems and increase developmental outcomes while still producing sufficient energy to meet demand. So we could refocus on community-based mini, micro and even pico hydro. That is a view shared by many environmentalists, but it does conflict to some extent with the role that hydro, and pumped hydro storage projects with large reservoirs especially, might play in helping to balance variable renewables like wind and solar PV.

Even though, once built, large hydro projects can produce relatively low-cost power, given the environmental problems and social impacts environmental groups have often strongly resisted new large hydro projects. It is true that wind and solar PV have now reached the point when their capacity combined is roughly equal to that of hydro, and they are continuing to expand. However, hydro does supply around 3.6% of global final energy at present and, although some of that might be replaced in time, excluding hydro entirely from the future energy mix would be a significant step. It would leave a gap in many scenarios.

Nevertheless, opposition has continued. For example, the proposed Green New Deal in the United States has opened up a debate as to what options should be included, with lobbying from 626 organizations, mostly environmental groups, including 350.org and Greenpeace USA, calling for 100% renewable energy 'by 2035 or earlier', with nuclear, large hydro and biomass/waste combustion all ruled out, along with all fossil fuel extraction: 'in addition to excluding fossil fuels, any definition of renewable energy must also exclude all combustion-based power generation, nuclear, biomass energy, large-scale hydro and waste-to-energy technologies' (Climate Change letter 2019).

That was clearly a 'maximalist' position and it was not supported by some of the larger US environmental groups, including the Sierra Club, the Natural Resources Defense Council and the Environmental Defense Fund (Atkin 2019). Even so, there are many who oppose new large hydro (as well as nuclear and biomass) and some around the world who would like to see old hydro plants and dams generally removed on environmental and social-impact grounds (Breyer 2019). Interestingly, one recent study suggested that, in theory, solar power plants could replace all (mainland) US hydro dams' power output using 'just 13% of the space' taken by their reservoirs (Waldman et al. 2019). That is rather speculative, especially given the very different operating characteristics and grid-balancing capacities of solar and hydro but, if nothing else, it does indicate that those who want hydro dams removed are thinking about the energy implications.

As noted earlier, there is usually less opposition to smaller hydro projects (100 MW or less), which currently make up about 10% of global hydro capacity, and there is potential for significant expansion. For example, small hydro contributes about 3% to the total electricity generation in Europe, with over 17,800 small schemes and a total installed capacity of 12,333 MW in the EU-27, with considerable potential for more. There are also around 48,000 small hydro projects in China, many under 10 MW, and 90% of recent clean-development mechanism-supported projects in China have been for small hydropower. There are many sites still available globally, so we might be able to shift to mini-hydro and to smaller, less invasive, run-of-the-river schemes without reservoirs (Moran et al. 2018).

For the moment, hydro, large and small, is still retained to some extent in most future high-renewables scenarios but usually without much expansion. For example, it supplies 3% of global primary energy in the LUT/EWG 2050 global scenario, with a small rise in capacity from now, and 4% of total final end-use power globally in the Jacobson team's 2050 scenario, with no new hydro capacity added but higher load factors assumed. His more recent scenario, with overall energy demand lowered, raises its share to 5.1% (Jacobson 2019a).

However, while conventional hydro may not expand in scenarios like this, there are new, or additional, roles for hydro emerging, which some scenarios focus on more; it is now often seen as a key to balancing variable renewables via pumped storage systems. Hydro already provides some balancing, without the need for pumping: if a hydro plant's reservoirs are already topped up, it can produce extra power rapidly, on demand, when there is a shortfall on the grid, but pumped storage adds more flexibility and balancing potential. Surplus power output from wind and PV solar can be used to pump water uphill into the reservoir for later release to generate power again.

Globally, there is around 127 GW of hydro-pumped storage capacity, with expansion underway. Some say the potential for that is very large. A recent GIS-based study suggested that there were potential sites globally for 22 million GWh of hydro-pumped supply capacity, much more, if it could be linked up appropriately, than could ever be needed for grid balancing. Most of that would be non-river-based 'closed loop' systems, with water shifting between high and low reservoirs (Blakers et al. 2019). It will be interesting to see how much of that materializes and what its impacts will be.

Clearly, large conventional hydro has problems, including major dam-failure risks and eco-impacts. If hydro is to remain a major and expanding renewable source, better design and siting and more careful maintenance and monitoring will be vital to reduce failure risks and impacts. However, although there may be ways to limit the impact of new large projects, it could be that the scale of hydro projects will have to change and, as climate change-related water shortages make hydro less reliable as a base-load supplier, its role may increasingly be to support and complement other more variable renewables via pumped storage.

Some of hydro's environmental impacts are shared by tidal barrage since they dam off whole estuaries. That is not a problem with tidal lagoons, tidal current turbines or wave energy devices. Most high-renewables scenarios include a contribution from some of these options, although it is usually relatively small. For example, Jacobson and colleagues have earmarked just over 338 GW of wave and tidal capacity by 2050 globally (Jacobson et al. 2017), a vast increase from the well under 1 GW now in place but small compared to the potential. For example, IRENA has put the total global tidal energy resource potential (including for barrages and lagoons, as well as tidal current systems) at 1 TW and that for wave energy at maybe 3 TW. How much of that will ever be exploited remains to be seen but, on the basis of progress so far, tidal current turbines actually look likely to develop faster and achieve greater market penetration. Wave devices, although harder to develop, may well catch up at some point, and while large tidal barrages seem unlikely to prosper, most environmental groups being opposed to them, smaller ones might do so in some locations, as may tidal lagoons.

These new types of water-power use are mostly still at an early stage of development, but, as the potentials indicate, it is conceivable that combined wave and tidal could at some point eclipse hydro and, with the exception of large tidal barrages, they seem likely to have much lower environmental impacts than large hydro. Then again, some existing large hydro plant can be converted to pumped storage, as is being planned for the 2 GW Hoover Dam in the United States (Penn 2018). The hydro debate continues (Hodges 2019).

Yes to solar heat

Turning now from technologies like hydro, which might be opposed or revamped, to other technologies, in addition to wave and tidal, some of which could be expanded, it is worth noting that so far I have been looking just at electric *power* production. The heat side, however, is also important, though until recently, biomass apart, less well developed.

As indicated above, solar electric PV power prospects are looking good, but it is often less recognized that so too is solar heat. There was around 480 GW of PV solar installed by the end of 2018, and much more expected, but there was also around the same amount of direct solar thermal *heat* system capacity in use (over 470 GW of thermal), and that too is growing. Most of this capacity is in the form of standard rooftop solar heat collectors, with China in the lead (330 GW of thermal), but large community-scaled solar heating arrays have also been developed in Europe and elsewhere, some of them linked to large inter-seasonal heat stores, allowing summer heat to be used for winter warming via local district heating networks.

Denmark has been a leader in this field, with its flagship 13.5 MW Marstal project along with many others around the country (PlanEnergi 2019). The pond-type heat stores typically used involve large lined water-filled pits, with floating insulating covers for heat retention. In most cases, the solar input augments heat supplied by other means, including biomass combustion, but new approaches are being adopted which enhance the value of the solar heat input by using large heat pumps. Although (fossil) gas-fired heating still often has the edge, solar heating with heat stores can be near competitive with other heating sources if district heating (DH) networks already exist, as they do in Denmark. The balance may tip further in favour of solar/DH/heat store systems, given that they avoid the carbon emissions associated with using fossil gas for heating.

Pit-type water-filled heat stores are not the only option. In some locations, use is made of underground thermal energy storage, where excess heat is simply stored in the ground. The Drake Landing inter-seasonal solar heat storage system in Canada, where of course winters are very cold, uses deep vertical boreholes (Drake 2019). UK company ICAX is developing similar ideas with commercial-scale solar-fed inter-seasonal heat stores (ICAX 2019).

So one way or another, the idea of solar heating and storage is spreading, there being, besides Denmark, many district solar heat-augmented network projects around the EU, including in Austria and Germany. In parallel, a completely new solar technology, hybrid solar-thermal/PV ('PVT'), is being developed around the world. This avoids overheating the PV cells, which can reduce their efficiency, by extracting usable heat. In some variants, a layer of semi-translucent PV sits on a solar thermal back plate, although other systems use focused solar in evacuated tubes. Essentially, they are solar combined heat and power systems, getting higher overall energy-conversion efficiency and, by doubling up on PV and solar heat collection, making better use of roof space.

Another possible area of expansion is direct solar cooling. As climate change impacts, this is likely to become urgent. Air conditioners can of course be run using solar PV power, but it is also possible to run absorption chillers or evaporation devices using solar heat. There are some district cooling networks and they can be linked up to cool stores run off solar chillers/evaporators. Note that, whatever the energy source used, storing 'cold' is just as viable and valuable thermodynamically as storing heat and, as with heat storage, cold storage systems can provide a way to balance variable renewable inputs (including from solar) and variable heating/cooling demands.

Direct solar heating and cooling should continue to expand, along with more advanced hybrid PVT systems, and can aid power-grid and heat-supply balancing. However, further ahead, we might also look to direct hydrogen production using focused solar heat for the thermal dissociation of water molecules: conversion efficiencies are low but improving (Rao and Dey 2017).

A 100 kW hydrosol project, using CSP-type heliostat mirrors, has been running in Spain since 2008, and novel approaches to high-temperature dissociation have been explored elsewhere, some with grid-balancing potential. In the Hytricity system, still being developed, solar thermal-derived hydrogen is stored, ready to be used to generate power at night via a high-efficiency combined cycle turbine. Whereas with conventional CSP heat is stored to enable 24/7 power generation, this new system makes use of a hydrogen store, which may have efficiency advantages. It is claimed that this system could achieve an overall energy-from-sunlight capture efficiency of around 35%.

There are many other direct solar heat-based water-dissociation systems under development, some of them aiming to produce 'solar fuels', i.e. synthetic gases and liquids, including aircraft fuel, by using solar heat, water and air-captured carbon dioxide. It would certainly be interesting if it proved possible to make fuel just from sunshine, air and water.

As can be seen, in addition to its current wide-scale use for heating, there is a range of new ways in which solar heat can be used. Clearly, while PV solar electricity generation often gets much attention these days, the use of solar heat also has big potential for expansion with, in addition to direct solar heating and cooling, new applications emerging, including for fuel production, storage and balancing. Given that it is generally much easier to store heat than electricity, we are likely to see many more ideas for heat (and cold) storage, solar inputs being one obvious route forward, and some of these systems will be able to aid balancing. Storable solar fuels can also be used for that, as well as for many other purposes. Solar heat really does seem to offer quite a range of end-use possibilities.

The technology mix

In some ways, we are spoilt for choice with renewables. There is a large and expanding range of options. As yet, it is unclear exactly what the optimal and acceptable mix will be. As this chapter has indicated, some of the renewable supply options are less popular than others. Some types of biomass use may well be unacceptable and so may large hydro. Their exclusion or limitation could be problematic, including for balancing (both can provide firm power and storage), although there are other options for that and some of them could be substantial. For example, I have mentioned the idea of converting surplus power from wind and PV solar into heat or hydrogen and storing that for later use, including to use again for grid balancing. In addition, while new large hydro projects might be constrained and large tidal barrages are probably not viable on cost and eco-impact grounds, smaller tidal barrages and tidal lagoons might play a pumped-storage role. As the review above of solar heat options indicates, hybrid solar heat and power systems, including those with heat stores, could also play a growing role. So could geothermal heat and power. Indeed, some see the latter as expanding in some locations to become a major, and non-variable, renewable energy source (IRENA 2017b).

So there should be no shortage of power or balancing options, even if all hydro and biomass options are excluded. However, clearly if some of the latter are retained, as seems likely, more options become available. For example, if the use of bio-waste is accepted, then, as mentioned in chapter 2, biogas-fired combined heat and power plants, with heat stores, can help with flexible balancing by varying the ratio of heat to power out so as to match varying renewable power supply and varying heat and power demand.

We also need to consider the demand side much more seriously. Not only does demand vary, it is also increasing. However, as I noted in chapter 2, there are ways to deal with both issues. Indeed, smart grids, demand reduction, efficiency improvements and supply-side generation technologies can all work well together: we need to operate on both the supply and the demand sides, not least since flexible-demand management can help with balancing variable renewables by reducing peaks.

Tragically, demand reduction and energy efficiency are still often the poor relations, often symbolically mentioned but mostly marginalized in the focus on supply. As the IEA keeps insisting, that has to change: there are major economic and energy gains to be made from investment in efficiency (IEA 2018b). Hopefully, awareness of the importance of flexible-demand management will add extra impetus. Most of the scenarios looked at above do take that on board, although to varying degrees.

As I have noted above, the IEA and others have also been pushing CCS and negative emissions technologies (NETs) as a way ahead, this forming part of the 'net zero carbon' approach adopted by some governments. I reviewed some of the problems with CCS, in relation to biomass and BECCS, earlier in this chapter, and I will be returning to some of the technical and strategic issues in chapter 4 in relation to system balancing.

I did mention earlier one specific NET option, direct air capture (DAC), the chemical absorption of CO_2 direct from the air. That has recently been talked up quite strongly. DAC plants have the advantage over fossil CCS plants, in that, in theory, DAC plants can be set up anywhere: CO_2 comes from the air, not from power station exhausts. However, they are likely to be expensive. They need energy to run (which may in fact constrain where they can be sited, as may the need for CO_2 storage), and the proportion of CO_2 in the air is very low (only around 0.04%), so to capture a ton of CO_2 you have to chemically process over 2,500 tons of air. By contrast, the CO_2 proportions in the outflows from fossil-fired power stations and some industrial processes are very much higher (3–15% for power stations, depending on the fuel and the technology), giving fossil CCS a major advantage over air capture. It is conceivable that DAC plants could be run off renewables (for example, PV solar), but would it make sense to use green power for this carbon removal activity, rather than using it directly, so replacing the need for carbon dioxide-producing plants?

It has been argued that, however it is configured, the artificial carbon capture/carbon removal approach is not viable technically or environmentally as a long-term solution to climate problems, not least because of the vast storage space for CO_2 that would be needed (Boysen et al. 2017). On this view, at best this approach provides a short-term palliative, at worst a deflection from the rapid development and deployment of zero direct emissions renewables (Elliott 2019d).

It is also argued that the 'net zero carbon' model, with carbon removal and renewables combined as if they were of equal long-term merit, is flawed: some say there should be separate targets for each to avoid the risk of post-combustion carbon removal undermining the expansion of zero-carbon generation (McLaren 2019). That formulation does imply that carbon removal options may have at least an interim compensatory role, whereas most of the academic/NGO scenarios avoid or limit CCS, with, for example, Jackson and colleagues excluding fossil CCS and BECCS. LUT, while critical of BECCS, does however see DAC powered with PV as offering significant carbon removal possibilities in some desert locations (Fasihi, Efimova and Breyer 2019).

Some scenarios do also look to *natural* carbon removal options and to protecting or expanding carbon sinks. Certainly, planting more trees and enhancing natural soil sequestration are important for a range of environmental reasons but, in carbon terms, these options are still just ways to compensate for past and continued use of fossil fuels. We cannot expect to keep on doing that indefinitely, finding room and storage opportunities to deal with ever-increasing emissions.

It does seem clear that, in addition to improved energy efficiency, avoiding emissions at source by switching to renewables is the preferable approach, especially in the longer term. In general, most of the 'high renewables' scenarios assume that electrification will spread and that much of the energy for heat and transport will be provided by renewable electricity. In part, that is because that may be an easier route than trying to decarbonize heat and transport using green energy sources direct, although the direct use of solar heat, biomass and geothermal energy can play a role. For example, some NGO scenarios include a significant element of local green energy generation for a range of purposes, such as biogas-fired CHP for heating networks (CAT 2013; RTP 2015).

Overall, specific end uses apart, it seems clear that having a range of renewables in the mix will be important for balancing and system security: that can smooth out and compensate for the variable outputs from some of them to some extent. For example, there is a rough inverse correlation between seasonal availability of power from wind and from the sun. However, given that not all the variable supply inputs will be conveniently matched (e.g. there can be periods of no sun and no wind), balancing is going to be increasingly important as more renewables come on the grid. Indeed, some see the variability of renewables like wind and solar as a key constraint to their wide-scale use. The reality is that, as argued above, even if biomass and hydro are limited, there are many ways to deal with this. As I explore in the next chapter, some involve new infrastructure and upgraded system management, but some involve the integrated use of renewable heat, power and storage.

What we are seeing here are options for the integration of a range of supply technologies into wider storage and balancing systems. That is central to most of the 'high renewables' scenarios I have looked at: it is not just about individual supply technologies making individual power inputs. They can and must also do other things. For example, in Jacobson's scenario, CSP, with integral heat stores, plays a role in balancing, as well as in power supply. In the LUT/EWG scenario, energy storage, fed by a range of renewables, can meet nearly 23% of electricity demand and approximately

26% of heat demand, with some of that coming from solar heat, which overall supplies 5% of global primary energy. In addition, electrolytic conversion of surplus wind and PV power to hydrogen gas ('power-to-gas') plays a role in many 'high renewables' scenarios, meeting a range of end uses and also helping with balancing.

Many of the emerging energy-conversion options cross conventional fuel type/energy boundaries, sometimes reversibly. For example, electrolysis converts power to gas, fuel cells convert gas to power and these functions can be combined in hybrid units, offering more system flexibility. There are also so-called 'industrial ecology' opportunities for cross-sector integration and cooperation, for example with energy or material waste outputs from one plant being an input for another plant. So there are multiple possibilities. In the next chapter, I look at the system integration options in detail. Not all of the cross-boundary options will be economically competitive, given the efficiency losses of multiple energy-conversion stages but, as the next chapter shows, there is a range of flexible options for integrated energy system balancing.

4

System Development: Tying It All Together

The emerging renewable energy-based system will need to be integrated and optimized so that it can deal efficiently with and balance variable energy supply and energy demand. This chapter looks at how that might be done with a mix of energy storage and energy transmission systems for power but also for heat and gas.

Grid balancing and system integration

So far, I have been considering individual renewable energy supply technologies mostly in isolation. However, given that many of them produce variable outputs, they have to be integrated with flexible grid-balancing systems to provide a reliable source of power.

As I noted in chapter 2, when considering balancing options beyond simple backup plants, the usual starting point is a focus on battery storage. Batteries can store renewably generated power for later use when less is available or demand is higher, with batteries being seen as an ideal match to domestic PV solar and also enabling a switch to electric vehicles.

With Li ion battery costs falling by nearly 85% since 2010, the use of battery systems is certainly growing for all uses and at all scales, with a massive 55% growth per annum expected for Li ion batteries up to 2022 (Spector 2018). In its 2018 'New Energy Outlook' report, Bloomberg predicted that the overall global market for battery storage would be worth $548 billion by 2050 (BNEF 2018a). In its 2019 review, BNEF upgraded that to $662 billion, with over 1 TW of storage in place by 2040, led by utility-scale projects. Box 4.1 looks at some of the technical options and the prospects for development and expansion.

Box 4.1 Battery storage – options and prospects

Although relatively low-cost lead–acid batteries are still widely used, more expensive but higher-performance lithium ion batteries are currently in the lead for most applications, including domestic PV backup and utility-scale grid support, as well as for electric vehicles. The market for the latter initially provided the main boost for battery development. However, although Li ion batteries still dominate, other battery technologies are emerging for some uses, including, at the utility scale, high-temperature sodium-sulphur (NaS) batteries and flow batteries, which mix separate chemical electrolytes to create a charge in a reversible process.

Whatever type, most batteries can only be used economically to store power for relatively short periods, hours or at most a day or so. So, for example, in the utility context they are used briefly to meet temporary shortfalls in supply, and also as a buffer, to help maintain system stability. They are of little use for longer-term storage for meeting long lulls in power availability, and even the largest ones so far (e.g. some large Li ion projects and more advanced flow battery schemes), designed for high-power delivery, only have total energy delivery capacities of a few hundred MWh. What that means is that, for example, if suitably sized, a 100 MWh-rated unit could deliver 100 MW for one hour.

In the domestic context, smaller-scale/-capacity units are used (a few kW/kWh), but if suitably sized a battery pack can store some of the output from a rooftop PV array for use at night, assuming there has been enough sunlight available during the day. However, if that is not the case, or if skies remain cloudy for a while, the batteries would be no use and other sources of power would have to be used. From the consumer's point of view, that could be a problem (battery packs are expensive), but, from the power system's point of view, when suitably charged, aggregated and linked to the grid, domestic batteries can play useful roles in short-term grid balancing by supplying extra power to help meet demand peaks, which typically only last an hour or two.

Given that the availability of a distributed network of storage batteries may help to reduce short-term demand on the grid, for example at peak demand periods, batteries are seen as potentially important for grid balancing and support. However, as Box 4.1 notes, there are limits. Batteries have relatively low storage capacities and can only provide energy for relatively short periods, unless you have a lot of them. They are fine for a few hours, or maybe days, but cannot function for weeks or months. In fact, not many of the storage systems we have at present can. Even large pumped hydro can only provide power for a day or so at most, depending on the size of the reservoirs.

So are there other options, for longer-term bulk storage, to deal with longer-term lulls in renewable availability? Storing heat is much easier than storing electricity. As noted in chapter 3, heat can be stored over long periods, even inter-seasonally, with low energy losses. So can gases, such as methane or hydrogen, as well as compressed air, and the potential for their use as bulk-energy storage media in underground caverns is very large (see Box 4.2).

Box 4.2 Gas storage potentials

Natural gas is already stored in bulk in the EU, the US and elsewhere, but, looking to the future, the Energy Technologies Institute has said that there are tens of GW equivalent salt cavern sites in the United Kingdom (Gammer 2015), and similar capacities have been identified elsewhere, some of which could be used for hydrogen storage with low costs (Londe 2018). The hydrogen option is explored in more detail in the main text below.

Large-scale compressed-air storage is another option and is being looked at worldwide. A 2014 study saw compressed-air energy storage systems (CAES) as offering good performance, long lifetime, low net environmental impact and reasonable cost compared to rechargeable batteries (Luo et al. 2014). More recently, a joint Edinburgh/Strathclyde University modelling study suggested that wind-derived power could be used to compress air to be stored in porous sandstone strata offshore during the summer, ready for use to generate power again in the winter, with a round-trip energy efficiency of 54–59%. However, it would be expensive, at the very least doubling the cost of electricity (Mouli-Castillo 2019).

So far, although some small projects exist, there are no large-scale green energy-based gas storage systems in use, but some are planned. For example, underground compressed-air and hydrogen gas storage feature as part of an ambitious 1 GW renewable energy project planned in Utah, making use of a very large salt dome cavern (Weaver 2019).

Storage of gases allows us to balance supply and demand variation *temporally*, providing power *when* needed, but gases can also be transmitted to deal with balancing *spatially*, proving power *where* needed. This can be done over long distances with much lower energy losses than with electricity transmission. As we move to develop balancing systems, non-electrical storage and transmission options like this could play a major role. In fact, in part, they already do. At present, as I have noted before, most grid balancing is achieved by using gas plants that ramp up and down in response to changing demand and changing renewable (or other) power inputs. Their fuel is delivered by pipe and then often stored ready for use to make power for power-grid balancing when needed. You could say that gas storage *before* combustion is the best type of storage: it is easier to store gas than to store the electricity produced by its combustion.

Moreover, it does not have to be fossil gas. As I have already indicated, we can shift to biogas or other green gases, including hydrogen made using surplus power from renewables. The so-called 'power-to-gas' (P2G) electrolytic technology needed for this is developing rapidly. That is perhaps not surprising since it offers a way to turn the 'problem' of renewable variability into a solution by converting the surplus renewable power that will be generated at times into gas and storing it ready for generating power again, possibly in fuel cells, when renewable availability is low and demand high (Lambert 2018).

The pros and cons of P2G are much debated. In addition to playing a role in grid balancing, it can be used to supply hydrogen for other uses, including heating (via the gas mains) and as a vehicle fuel, used either in combustion engines or to power on-board fuel cells in electric vehicles. The latter two applications (mains gas injection and vehicle fuel production) have in fact been the main focus for P2G so far, for example in Germany, where there are many P2G projects underway. So hydrogen may have multiple uses, and some see it becoming a major new energy vector, not just for balancing.

The hydrogen economy

Hydrogen is not a renewable resource as such; it has to be made using an energy source. To be viable for wide-scale and environmentally sustainable use, 'green hydrogen' from P2G electrolysis does have to get cheaper. At present, making hydrogen this way is more expensive than by the more conventional route of high-temperature steam reformation of fossil gas (SMR), producing what is sometimes called 'brown' hydrogen. However, unlike P2G electrolysis, the SMR process generates carbon dioxide. Capturing and storing that to try to make the process carbon-neutral would add significant extra cost.

Fortunately, although there are uncertainties as to future trends, green hydrogen production by P2G electrolysis does seem to be improving, given lower costs and higher efficiencies. For example, global energy consultants DNV GL say that cheaper electrolysers, more frequent periods of low- or zero-cost renewable-produced electricity and penalties for carbon emissions will make the use of hydrogen produced from surplus renewable electricity competitive with natural gas-derived hydrogen before 2035 (Richard 2018a).

Others have declared that might be achieved by 2030 or even earlier in some markets. One study claimed that P2G hydrogen production, using excess green power, was already competitive in niche markets and could be so across the board in ten years (Timperley 2019). The IEA says that 'the cost of producing hydrogen from renewable electricity could fall 30% by 2030 as a result of declining costs of renewables and the scaling up of hydrogen production' (IEA 2019c). A study from Stanford University was also quite optimistic about the economics of renewable P2G hydrogen production and storage (Hanley 2019).

Moreover, some do not want to use just *surpluses* and claim that better results could be obtained if renewable power output was dedicated full time to producing P2G hydrogen. For example, consultants Navigant say that, 'once all the demand for direct electricity is satisfied with renewable energy, you can build additional wind turbines and solar panels specifically dedicated to producing green hydrogen' (Navigant 2019).

Interestingly, in this context, it has been claimed that hydrogen production using power from offshore wind turbines could soon be commercially viable (Bedeschi 2019), and there has been a UK proposal for a 4 GW floating offshore wind farm for *on-board* P2G hydrogen production, with the gas being piped to shore. The developer asserts, 'if you had 30 of those in the North Sea you could totally replace the natural gas requirement for the whole country, and be totally self-sufficient with hydrogen' (Lee 2019).

So, one way or another, we may see P2G hydrogen becoming a major option for grid balancing, being fed to gas turbines or possibly to fuel cells for backup power generation. It could also be used direct for heating and in vehicles, assuming there was sufficient surplus green power available to make enough of it for multiple purposes. The fact that surplus hydrogen production can be stored makes that more credible.

What about hydrogen *transmission* – getting it to where it is needed? As noted earlier, transmitting energy by gas pipeline can be superior to transmitting electricity by wires since the energy losses are less. In addition, the pipes can be put underground, something that is much more expensive for high-voltage AC electric cables.

At present, in the United Kingdom, about four times more energy is shifted by gas pipes than by the power grid, and the gas grid also provides storage capacity. Some say that hydrogen gas could replace the fossil gas currently used. It could be used just like methane for heating and cooking after a few simple modifications to domestic appliances. Burning it produces water vapour but it also leads to nitrogen oxide (NOx) formation, like all combustion.

However, there are what may be more substantial issues in relation to possible leaks. Gas grids can and do leak (for example, more than 2% of US fossil gas production is lost to leakage), and releasing methane into the air is a major climate issue. It is a much more powerful greenhouse gas (GHG) than carbon dioxide, with a half-life of about seven years in the atmosphere before it is oxidized to CO_2 and adds to the latter's accumulation. Although much lower than that of methane, hydrogen does have a GHG impact as well as other possible problems. It is a smaller molecule (H_2) than methane (CH_4) and can find its way through cracks and seals relatively easily, and can also lead to the embrittlement of metal pipes.

These issues are being looked at in the United Kingdom. It has replaced most of its old iron gas pipes with plastic pipes, though not yet all of them. So at present the gas grid system may not be able to safely handle even medium levels of hydrogen mixed with methane (e.g. a 20% mix), much less 100% hydrogen. However, that will change. The UK Institution of Engineering and Technology claims that, when the gas grid has been fully upgraded, full conversion to 100% hydrogen delivery should be viable in safety terms (IET 2019). Nevertheless, the leaks issue will need to be watched. A UK government short review maintains that the use of hydrogen for heating will have low impact, but it admits that more research is needed (BEIS 2018a).

As I have indicated, there are multiple possible uses for hydrogen, not just heating and balancing. Looking globally, IRENA also notes that important synergies exist between hydrogen and renewable energy. It can increase renewable electricity market-growth potentials substantially and broaden the reach of renewable solutions, for example in industry. It adds, 'Electrolysers can add demand-side flexibility. For example, European countries such as the Netherlands and Germany are facing future electrification limits in end-use sectors that can be overcome with hydrogen. Hydrogen can also be used for seasonal energy storage' (IRENA 2019b).

Some say that it might be best just to generate and use hydrogen locally, perhaps using local renewable power sources with small local hydrogen gas stores, so as to avoid long-distance hydrogen transmission. However, if as seems likely the gas transmission issue can be resolved, it might make more sense to have large high-efficiency electrolysers near large renewable energy sources, for example in the UK context, at port towns currently used to land imported gas (LNG), which have good gas grid links to energy demand centres but also good access to offshore wind farms. Similar options exist elsewhere, along with the possibility of long-distance tanking of P2G hydrogen (or ammonia made from it) by ship from locations where renewable power supplies are plentiful. So it could be that the use of hydrogen, or derivatives, and its transmission or transport over long distances could become widespread.

Supergrids

The (gas) pipe versus (electric) wire energy transmission debate still continues. Long-distance pipe links already exist for fossil gas transfer, for example from Russia to Europe, and gas is shifted by tanker ships around the world. Gas tankers may have their limits (they need fuel for propulsion) but, as noted above, gas pipes, including for hydrogen transmission, can be preferable to conventional high-voltage alternating current (HVAC) power transmission in some locations. However, it seems unlikely that, in most places, gas transmission will ever fully replace power grids for very long-distance energy transfer, not least given the advent of more efficient high-voltage direct current (HVDC) 'supergrid' power transmission. They can cut energy transmission losses to 2%/1,000 km, compared to maybe 10% for conventional HVAC transmission and distribution, primarily because there is less heat loss on the cable with HVDC.

That opens up a new possibility. The lower heat loss makes it easier and less costly to bury HVDC links underground than those of HVAC. It is cheaper to have cables on towers for either option, with the heat being dissipated in the air, but, although still very expensive to do, burial is cheaper using HVDC. Burying the wires would not only reduce objections to the visual intrusion of power lines and towers in sensitive areas but might also help reduce the risk of accidental wildfires, which are sometimes caused by (tree) contact with, or failure of, overhead cables.

There are technical issues with HVDC. It is harder to upload and download power to and from HVDC grids than to and from HVAC grids. With AC, use can be made of transformers to convert voltages relatively easily, but with DC use has to be made of more expensive switchgear and conversion systems. However, in the context of long-distance HVDC transmission, power can just be uploaded and downloaded at each end, limiting the use of this equipment. AC grids make more sense for shorter-distance local distribution, with easier multiple, intermediate-location downloads or (in a system with distributed generation) local uploads. So the two can work together: it's horses for courses.

Connection specifics aside, where long-distance power grid systems come into their own is in their ability to enable the near instantaneous balancing of variable supply and demand. Gas systems can have large gas stores to allow output to be ramped up fairly quickly and the gas pipes themselves act as gas stores, but power grids can provide very flexible and fast system integration. Suitably sized and configured, they can be used to allow excess outputs in one location to meet demand in another, so helping to balance the local variability of renewable sources. As I noted in chapter 2, supergrids can give access to pumped hydro storage if it is not available locally, as well as, crucially, to sources over a wider geographical area, which increases the probability that there will be surplus power available for trading. The wider the geographical spread, the better the balancing potential (Fraunhofer 2016).

For example, long-distance HVDC supergrid networks could be used to shift balancing power between projects and countries in the EU and also to link to desert solar projects in North Africa, feeding power to the EU, as I explore in chapter 6. However, as I discuss there and have also looked at elsewhere (Elliott 2013a), there could be exploitation and equity problems with that type of trade and, so far, although there have been proposals for major schemes, only one such project is still being explored seriously (Casey 2017). Nevertheless, the very large solar resource in North Africa is unlikely to be ignored for long.

Meanwhile, given that it is technically possible to trade power over long distances, other ambitious power-balancing and power-trading projects are likely to emerge. At present, most countries do some cross-border power trading, but there have been proposals to extend links between national power grids across the EU and also across Asia, as in the so-called 'Golden Ring' pan-Asian supergrid (Movellan 2016). That clearly opens up some interesting geopolitical issues, as I will be discussing in chapter 6. So do the even more ambitious proposals for links between continents, for example between North America and Europe (Purvins et al. 2018).

Going even further, China, which is already building a giant HVDC network internally, has reportedly also been looking at the idea of a *global* grid (Baculinao 2016). That may be fanciful but, if it were ever achieved, amongst other things it would enable solar power from the sunlit side of the planet to be used by those on the night side, the ultimate in balancing options.

Local power smart grid trading

It may be a long way from science fiction-sounding ideas like global power trading and balancing to the current everyday reality, but the power-trading system is changing rapidly at all levels, as can be seen by the advent of new 'smart grid' local trading systems. They allow local self-generating domestic 'prosumers' to trade surplus power on a peer-to-peer basis. As noted earlier, generation by prosumers and local energy co-ops is expanding in parts of Europe and also elsewhere, with local mini-grids also emerging. Local smart power-trading systems can help them spread and integrate into more effective and efficient power systems (see Box 4.3).

Box 4.3 Smart power-trading systems

There is much enthusiasm for smart interactive supply-and-demand management systems at the local level, since it is claimed that they can reduce costs by more efficiently matching and balancing local energy supply and demand,

and also by integrating local storage into the system, with a new local energy market emerging. For example, the new Smart Export Guarantee system deployed in 2020 in the United Kingdom aims to create 'a whole new market, encouraging suppliers to competitively bid for this electricity, giving exporters the best market price while providing the local grid with more clean, green energy, unlocking greater choice and control for solar households over buying and selling their electricity'. It is claimed that it 'would mean households and businesses installing new renewable energy generators would be paid transparently for the energy they produce – protecting consumers from cost burdens, by using established smart technology' (BEIS 2019c).

There are some issues with this type of market-based approach. As currently conceived, the local markets are created and run centrally by, and some fear for, the national power utilities. Decentralists may prefer direct peer-to-peer trading without too much external mediation. There are also technical issues. The roll-out of smart meters in the United Kingdom has been slow and problematic and, for wider system use, although some look to blockchain encrypted digital ledger systems – for example, for managing peer-to-peer power trading amongst prosumers – the blockchain system can require significant amounts of power to run. Hopefully, less energy-intensive versions will be developed: it would be counterproductive for the smart green power system to use much of the energy it generates just to support blockchain operation (Ahl et al. 2019; G. Brown 2018; WEC 2017).

As a perhaps eccentric technical footnote, it is worth pointing out that most modern electronic systems (TVs, PCs and so on) and some lighting systems run on 12 volts DC or similar, not high-voltage high-quality frequency-stabilized AC. The consumer electronic units each have their own small transformer voltage step-down and AC-to-DC rectifying system. That is wasteful in energy terms and can be part of the so-called 'vampire load' problem, since, at the home level, the individual mini-transformers in each device are often left on, even when the device is not being used. That can include chargers for portable devices. They generate a small amount of waste heat, very inefficiently: it is a poor way to heat a house! Can we do better? Do we actually need AC? AC power is needed for the type of electric motors currently used in heavier domestic equipment (such as washing machines) and also in industry, but there might be some value in having a separate low-voltage

DC grid for some uses. That could offer some interesting local low-power network trading opportunities, including for local PV power.

In theory, the resultant trading can help with balancing local supply and demand variations, though it may need links with the wider grid system to achieve full balancing. Nevertheless, the advent of smart grids with digital controls can help with integrated system power management, optimizing the value of local generation and storage and linking into flexible-demand management systems.

Demand-side management (DSM), sometimes also called 'demand-side response' (DSR), can be achieved in a number of ways. For example, in a simple version, if there are variable time-of-use power tariffs in place, power users may elect to avoid using selected appliances at high-cost peak energy demand times. Certainly, most fridge/freezers can happily coast for an hour or so with no power input, and the use of some other domestic appliances (e.g. washing machines, tumble dryers) can sometimes be delayed without inconvenience.

In a more sophisticated version of this demand management approach, if users so choose, and with override options, supply decoupling for selected devices can be done automatically in response to 'smart grid' signals when power costs are high. This does not just apply to domestic energy users. Many retail outlets, such as supermarkets, and some industrial plants can reschedule some types of power use and thus save energy and money (Ambrose 2019).

From the power system's point of view, the result is a shift in peak demand to times when, for example, renewable supplies are more available. It is a power-balancing system, which avoids having to build extra capacity to meet peak demand. So it saves money all round. So of course does investment in energy saving. Some of that can reduce not just demand overall but also peak demand, as can changes in energy use patterns adopted by consumers. Indeed, it is often hard to separate out the impacts of some of the interactions between new patterns of consumer energy use, energy-efficiency measures and demand management activities. I will be looking at issues like that in chapter 5.

In some variants of the smart grid supply and demand management approach, involving the optimal use of energy storage, the interactions can get even more complex. For example, it may be that, in addition to having domestic battery energy stores linked to their home PV arrays, many people will soon have electric vehicles, each with its own set of batteries. They will usually be linked to the grid for charging at night, but at times it may be possible and acceptable to extract power from them to meet grid-demand peaks in so-called vehicle-to-grid (V2G) mode. The logistics may be a little tricky. No one will want to find their car battery drained when they come to use it. However, with suitable drain-protection safeguards, contractual agreements and of course payments, some see V2G providing a useful new resource for grid balancing, vehicle batteries providing substantial distributed energy storage capacity (IRENA 2017c; Lance et al. 2019).

Nuclear for balancing?

Some also see nuclear power plants as providing balancing support for variable renewables, a claim sometimes made by nuclear proponents seeking to compensate for the poor economics of nuclear power generation. At first glance, this sounds reasonable enough. Nuclear plant output is usually steady and non-varying.

However, that is actually a problem. The fixed output from a large nuclear plant is not what is needed to balance varying renewable outputs. Nuclear plant power output to the grid can be varied to some extent by (wastefully) dumping steam, as with any steam-raising thermal power plant, but the basic nuclear heat-generation process can only be varied relatively slowly and not too often. That is mainly because radioactive xenon gas is produced when nuclear fission levels fall, and time is needed for this to disperse. Otherwise, it can interfere with safe operation. So while nuclear plants can reduce their reactor core power output slowly, for example at night when consumer power demand falls, they cannot easily or safely ramp power output up and down regularly and rapidly to meet the frequent and sometimes large and sudden variations in renewable outputs. It is conceivable that nuclear plants could mostly be kept running at low to medium power and then be run up to full power to meet occasional long lulls in renewable availability, though that would be an expensive way to run them. They are usually run at full power continually whenever possible to recoup their high capital costs (Morris 2018).

New types of smaller nuclear plants may be developed which are more flexible, with high-temperature molten fluoride salt thorium-based reactors being talked up as one way ahead, but that is some way off. Working with corrosive, radioactive molten salts at high temperatures will certainly involve rather tricky plumbing, with their likely costs, security and safety still being uncertain (Thomas et al. 2019).

For the moment, if we want to use power plants to balance short- and long-term renewable output variations, it is much easier and cheaper to use flexible gas-fired turbines, as is already widely done. In time, they can be run on biogas or other green gases to avoid carbon emissions and, as I have indicated, there are also several other balancing options. So it would not seem to make much sense to build expensive nuclear plant to try to balance renewables (see Box 4.4).

Box 4.4 Nuclear vs renewables: a UK example

The potential conflicts between nuclear and renewables can be seen from the emerging UK situation. Peak electricity demand in the United Kingdom reaches around 60 GW in the winter, falling to 20 GW in the summer. It currently has over 40 GW of renewables in place and may soon have around 60 GW. Although not all of that will be delivering power all the time, there will be occasions when it can meet all power demand. So what happens then to the 20 GW or so of nuclear that the UK government would like to see installed? Indeed, some have envisaged even more. Given the high capital cost of this nuclear capacity, it would make no sense economically to curtail its output at times of power surplus, and that would in any case be hard to do with inflexible nuclear plants, but equally it would make no sense to dump the output from the low marginal cost renewables. In the absence of substantial storage or export options, inflexible nuclear and variable renewables do not fit together well on the same grid.

There may be other options. Given that nuclear plants seem to be having problems competing in the bulk-supply electricity market, interest has been shown in using nuclear plants for other, possibly more lucrative, purposes. For example, they might supply heat some of the time by tapping steam off from their power turbines, running in CHP mode, for district heating networks or industrial use. Or they might produce hydrogen, either directly (by high-temperature dissociation of water) or, more likely, via electrolysis using nuclear-generated electricity, in P2G mode.

The nuclear CHP/district heating option is limited by the fact that, although heat can be transmitted short distances (a few kilometres) reasonably efficiently, it makes more sense to locate heat-generating plants near to population centres. However, given the safety and security issues, few people want to have nuclear plants of whatever size sited in or near cities, where the big heat loads are concentrated. The nuclear electrolytic-hydrogen option might be possible for part of the output from old plants, allowing them to be more flexible: they could cut back on the power used for this, and feed the grid, when there was a long lull in renewable power availability. It is not clear, though, if switching between these roles regularly to help with balancing would be economic. It is also not clear that it would make economic sense to build new nuclear plants just for hydrogen production, much less to run them flexibly for balancing.

Views on all of that differ, with the nuclear industry evidently being keen to find a new role for nuclear, hydrogen production being one option (IAEA 2019). However, with renewables expanding, there is likely to be no shortage of surplus power for low-cost hydrogen production. It is conceivable that new nuclear technologies will emerge in time, and some might find roles in supplying heat, power or hydrogen direct to industrial complexes. However, there seem to be limited prospects in terms of grid balancing, given that there are arguably easier and cheaper options.

Carbon capture to the rescue?

Nuclear plants may not be very relevant for balancing but, as I have indicated, gas-fired plants can be and already are widely used for this purpose. While some look to switching them to the use of green gas to avoid carbon emissions, others look to achieving that by adding carbon capture and storage (CCS) systems. Indeed, as noted in chapter 3, some have seen gas-fired plants, and possibly also coal-fired plants, with CCS added to them as a way of allowing the continued use of fossil fuel as a major power source, not just for balancing.

In whichever role they play, the CO_2 they produce will be captured chemically and then sent for indefinite storage in underground strata, including old empty gas and oil wells. However, there are problems with CCS: it looks likely to be expensive and may also be inefficient, possibly making its use with power generation plants unviable at large scale (see Box 4.5).

Box 4.5 CCS potentials and problems

Carbon capture and storage has been pushed strongly as a reduction option. At one stage, the IEA argued that, globally, 'CCS could deliver 13% of the cumulative emissions reductions needed by 2050 to limit the global increase in temperature to 2°C' (IEA 2015) and the UK Energy Technologies Institute backed the idea (ETI 2017).

However, progress has been very slow. Two large coal CCS projects are running in North America but, with concerns about the cost, the United Kingdom halted its £1 billion CCS competition in 2015. Norway also pulled out of CCS work and, in 2017, the US abandoned its flagship Kemper coal CCS project after massive cost overruns (to $7.5 billion); it has been converted to a conventional gas plant.

There certainly are problems. Although it is claimed that CCS can capture and store high percentages of the CO_2 produced by power stations and some industrial plants, it cannot easily do the same for vehicles. Moreover, the high capture levels claimed for power plant emissions, sometimes 90% or more, may not be achieved in practice. For example, it will take energy to run the CO_2 capture, transmission and compression/injection process. Taking the carbon debt of providing that energy from fossil sources into account reduces the net CO_2 capture level to maybe 60–70%. Once injected into storage wells/strata, some of the CO_2 may also leak out. Hopefully, sudden massive and potentially catastrophic leaks of suffocating gas can be avoided by careful site management and selection of geological sites, but some CO_2 may still leak, undermining the aim of the exercise. Finding suitably secure sites for the indefinite mass storage of CO_2 on an ever-increasing scale would certainly be challenging.

Given that the focus in this chapter is on balancing, this is not the place to explore the pros and cons of CCS in any more detail, especially since it may not make too much operational or economic sense to add expensive CCS systems to fossil gas plants that are used to balance variable renewables. The main attraction of gas turbines is that they are relatively simple, cheap and flexible. With CCS added, flexibility may be reduced, while the CCS system may not be used optimally.

Of late, as I noted earlier, more interest has been shown in carbon capture and *ultilization* (CCU) since that would avoid the need for storage and could yield potentially valuable synthetic fuels, although making them would require hydrogen. We are back to the P2G electrolysis option, using renewable sources to make 'green' hydrogen. That is clearly possible, and there may be some industries for which CCU would be helpful as a way to decarbonize, while potentially creating a new income stream, assuming that the new fuels that could be made (e.g. for vehicle or industrial use) are more valuable than the green hydrogen.

However, as with CCS plants, using fossil CCU plants for balancing may make their combined operation less economically viable and, unless the new fuels were stored ready for use in power plants to meet lulls in renewable availability (perhaps not the best use for this fuel), CCU may not have much of a role in grid balancing. The same could be said for direct air capture (DAC), extracting CO_2 directly from the air by chemical absorption and then storing it (variously being labelled 'DACS' or 'DACCS'). Some see that as playing a key negative carbon CCS role. It does have its attractions but there are also some issues, as I noted in chapter 3, quite apart from the problems with CCS.

In relation to our focus here on grid balancing, unlike fossil plants with CCS/CCU, DAC plants would not produce any energy (they *consume* energy), so they could not help at all with balancing, except perhaps by being turned off when energy demand was high. However, DAC plants could play a CCU role (hence 'DACCU'), if hydrogen was available, producing fuels for other uses, as with fossil CCU plants. In theory, as noted above, that fuel could be used for balancing, but in both the DACCU and the fossil CCU case, if green power and green hydrogen are available, it would probably be easier and more economic just to use them directly for balancing power or other purposes.

Another option is BECCS, the generation of energy from biomass combustion and the capture and storage of the resultant CO_2. As with fossil CCU and DACCU, the CO_2 might also be converted to new fuels if hydrogen is available, i.e. in 'BECCU' mode. However, while BECCS or BECCU plants could perhaps be run flexibly, with their power outputs varied to help with balancing as with fossil CCS and CCU plants, that might not make too much sense: the expensive CCS/CCU systems would then not be used optimally. Alternatively, synfuels from BEECU could in theory be used for balancing, although, as with fossil CCU and DACCU, that might not be the best use of them or of the green hydrogen that would be needed to make them.

In general, then, even if they can be developed successfully, these various carbon capture/processing activities do not seem likely to help much with balancing variable renewables. Some hybrid options are possible, with carbon utilization being backed up by storage of any CO_2 that could not be used. However, whatever the package, the various CCS/CCU options, sometimes now lumped together as 'CCUS', are essentially about compensating for continued fossil fuel use, while possibly making new fuels or, in the case of DACCS and BECCS, reducing CO_2 levels. There are debates about their role in these contexts (EACAS 2018) and in terms of the viability of 'net zero carbon' policies which rely on carbon removal (McLaren 2019). I will consider that further in chapter 5 and again in chapter 8, but in terms of my focus here on balancing, rather than adding CCS or CCU to fossil or biomass plants or using DAC plants, it would seem to make more sense just to use flexible gas-fired turbine plants and to start feeding them with zero-carbon green gas.

Energy mixes, balancing and integration

In this chapter, I have looked at a range of balancing options, some of which (nuclear, fossil or biomass CCUS, or DAC) seem to be of little merit for that purpose. Some of the others (including P2G hydrogen backup and other types of storage, along with smart grids and supergrids) have been included in some of the 'high renewables' scenarios I introduced in chapter 3. For example, in the LUT/EWG mix, in addition to batteries for short-term storage, for longer periods and overall balancing, use is made of 'power-to-gas' electrolytic conversion of surplus renewable output to storable hydrogen. Jacobson and colleagues also make some use of hydrogen, along with a mix of CSP with heat stores, batteries and pumped hydro storage. These scenarios, and some of the other 'high renewables' scenarios, also rely to varying degrees on improved grid links, for example with regional, national or even in some cases international supergrid power imports and exports.

However, some of these balancing options are relatively new, and debate continues about the role they can and should play in an integrated system. Some may be rivals. For example, if there is a lot of power-storage capacity in the system, then there may be less need for supergrid imports or for (green) gas-fired backup plants. If demand can be reduced and made more flexible, there may be less need for storage or supergrid links. If we shift over to a hydrogen economy and/or one in which heat is used as a balancing option, the pattern of trade-offs changes yet again.

In most of the 'high renewables' scenarios I have looked at, gas-fired backup plants play a major role in balancing, sometimes using green gas (biogas and P2G), but power storage, supergrid inputs and demand-side flexibility can reduce the need for that. So can the availability of power from non-variable renewables. Clearly, there is a range of options.

The optimal mix will depend on location and cost and on end-use requirements. I have explored some of these issues elsewhere (Elliott 2016) but, by way of a contemporary example, it is interesting to look at the recent UK debate over how to decarbonize heating. It was framed in terms of the relative merits of switching from the use of natural gas for domestic heating (fossil gas currently being the United Kingdom's main heating source) to either grid-delivered electricity or piped-in hydrogen. Balancing was implicit in the analysis and, although not explored directly, it is clear that each option would have different balancing implications, as well as implications for end use, energy storage, system integration and, crucially, energy distribution – via gas or power. In the event, what emerged was a compromise, not 'all wires' or 'all pipes', but a bit of both, power *and* gas (see Box 4.6).

Box 4.6 Distribution choices for green heat supply – a UK debate

In the United Kingdom, most domestic heat is provided by natural (fossil) gas, piped to homes by the gas mains. The two main decarbonization options that have been looked at are switching over to using electricity from the power grid to run heat pumps, and repurposing the gas mains to deliver hydrogen gas, made either via steam reformation of fossil gas coupled with CCS or by electrolysis powered with renewable electricity, i.e. P2G conversion.

One study concluded that the electrification approach, using heat pumps, would require a 67% increase in power generation, as well as grid reinforcement. It was noted that 'hydrogen-based heating puts less strain on the electricity system', although it still would require a 30% rise in power use to run the system (Aurora 2018). Another study concluded that the 'hydrogen-led heat decarbonisation pathway could be lower in cost by several tens of billion pounds' than an electrification-led or hybrid gas–electric approach (Element Energy and E4Tech 2018).

However, a study for the government's advisory Committee on Climate Change (CCC) said that a hybrid mix could be the least-cost option, with the 'hydrogen alone' route being seen as the most expensive (ICL 2018). In the event, the CCC recommended the hybrid approach. Most heating would be done with electric heat pumps, but they would be hybrid versions with integral gas boilers, which would allow peak demand to be dealt with by using mains gas and, possibly ultimately, hydrogen delivered by the gas mains (CCC 2018b). The government subsequently backed this plan.

This compromise may seem unwieldy, but it has the merit of limiting the strain that would be put on the power grid at peak heating times and also the high cost that the CCC felt would be incurred by switching totally to hydrogen production: they were not convinced that P2G would get cheap. Certainly, heat pumps do have the attraction of using electricity around three times more efficiently to produce heat than if it is used directly in conventional electric heating systems, and that means that using them would reduce the stress on the grid.

However, heat pumps still do use electricity, and the efficiency and reliability of some of the systems can vary with the weather. While there is certainly a role for domestic-scale heat pumps, especially in rural areas that are not on the gas grid, some argue that, in urban areas, larger heat pump units, supplying heat to district heating networks and heat stores, are a better bet in energy and reliability terms. Then again, combined heat and power (CHP) plants can do this even more efficiently. Whereas heat pumps can have a 'coefficient of performance' (COP) of 3, a large CHP plant can have a COP-equivalent of 9 or more, depending on the temperature requirement. Nevertheless, for the moment in the United Kingdom at least, domestic heat pumps and grid power seem to have (mostly) won the day, with hydrogen serving as a backup.

Similar pipe-versus-wire/power-versus-gas debates are underway elsewhere in Europe, where natural gas is also often the main heating source, especially in countries near to the (fast-depleting) North Sea gas fields, but also in countries where gas is imported from Russia. The situation in the United States is rather different. Although gas-fired 'peaker plants' are used for power-grid balancing, natural gas has not in the past played a major role in the United States in heating or power generation. However, the advent of large-scale shale gas extraction has changed that. Shale gas fracking may have significant environmental and health implications, but switching from the use of coal to the use of gas in more efficient gas turbines, which produce less carbon dioxide/kWh generated, has helped the United States to reduce its emissions.

That may be a short-lived gain. There may be limited prospects for continued substitution of coal by gas and also, possibly, for the continued use of fossil gas (Dyson 2019).

Reserves of shale gas (and shale oil) are finite, and shale gas well-production levels tend to fall off rapidly. Moreover, these are still fossil fuels so using them, for whatever purpose (power or heating), does still produce CO_2. Given environmentalist opposition to that, and also to gas pipelines, it is perhaps not surprising that the city of Berkeley in California recently decided to block gas use for heating in new buildings.

That opens up an interesting tactical issue. As noted above, there are methane leakage risks with pipelines and, inevitably, burning methane produces CO_2. However, burning hydrogen doesn't, and so the Berkeley plan to close the gas-supply system down for some uses, in preference to green power electrification, may be a short-sighted move. It would mean losing the option of using piped-in green hydrogen for home heating, a more flexible option that is now opening up in the United States, thanks to a new P2G system (Holliman 2019).

As the example in Box 4.6 illustrates, in the UK heating case, a trade-off was made between the rival pipe/wire options, seeking to avoid the ostensible limits of both. That is not easy to do. There will be limits to *any* approach to energy generation, distribution and balancing. For example, it would be hard to supply all the United Kingdom's heat demand with green electricity, peak demand especially. Although heat pumps are more efficient than most other heating systems, the power grid would have to be enlarged, and balancing the peaks might be difficult. Demand management and heat storage might help, but the size of the winter-evening heat demand peak is quite formidable in the United Kingdom and other similar locations. Hence the compromise that has been adopted in the United Kingdom.

There may also be limits on the hydrogen side. Unless the amount of renewable capacity installed is very large, there might not be sufficient surplus to make hydrogen for both heating and balancing, and also for other purposes. Then again, most 100% renewables scenarios typically have enough renewables to meet demand most of the time, so at times of low demand there should be plenty of surplus. Indeed, some have suggested that, since the cost of renewable power plants was falling, we could afford to install a very substantial *over-capacity* so as to be able to meet power demand at nearly all times, even at peaks, and that would generate even more surplus at other times. In one such proposal, it was claimed we could condone the curtailment of the large excess output at low-demand time (Perez and Rabago 2019). Carried to the extreme, curtailment on this scale seems to be a wasteful approach, not least since this surplus can be used for balancing, for example via P2G conversion and storage. That in turn might reduce the need for so much over-capacity. That seems a better option.

It is true that the use of green power surpluses (not continuous power) means that the expensive P2G electrolysers would be left idle part of the time, when there is no surplus, undermining their economics. As noted earlier in this chapter, an alternative option might be not to just use surpluses but to run the electrolysers continuously on green power, although there would still be problems, given the variable output of renewables. Another possible approach (mentioned in chapter 2) is to use some of the surplus power to warm up large heat stores, adding to heat from other sources, possibly from CHP plants, so as to meet heat demand peaks later on via heat networks. That might be more economic than using hydrogen from P2G for heating via the gas mains. However, it depends on whether conversion of green power to hydrogen by P2G electrolysis turns out to be viable on a significant scale and can be integrated into the wider system. Moreover, it might still be easiest just to use the green power directly, grid delivered, for heat and power, although then the storage and balancing option is lost.

System reliability, transport and social change

As can be seen, although the choices can be complex, there are many possible ways a 100% renewable heat and power system can be balanced so as to cope with the variability of renewables. However, there is another aspect of the new power system that needs attention. There may be problems in relation to system stability.

The existing power system has large power plants with massive rotating turbine generators. These provide rotational inertia (via their angular momentum), and the grid-linked reactive power helps to maintain system stability so that it all stays frequency lock-stepped. With the large conventional plants mostly gone under the new system, some form of replacement grid stabilization is needed. Some parts of existing power plant equipment can be retained for this purpose, acting like giant grid-linked flywheels, or new versions can be built, sometimes called synchronous condensers. However, it is also the case that some of the green energy systems can provide rotational inertia. Large CHP units, geothermal plants and biogas/P2G hydrogen-fired backup plants can do this, as can hydro plants. Wind turbines also offer some rotational inertia; so can wave and tidal turbines, but PV solar offers none. To compensate for that, battery-based inverter systems, with electronics to provide phased-frequency support, are being used, offering in effect synthetic inertia (Kroposki 2016).

So there do seem to be some 'technical fixes' available for this inertia problem, although they may add to the cost, and that opens up a more general system reliability issue. No power system can be 100% reliable. We can build in redundancy, balancing, synthetic inertia and backup reserves to minimize the risk of supply problems, but for all power systems there are limits to how much we can afford to spend on this. There can be diminishing returns so we set reliability criteria at failure levels deemed to be broadly acceptable.

That is done for all technologies and inevitably the risk levels chosen can be controversial. For example, much effort is expended on trying to limit the risk of major nuclear plant accidents, but they do still occur occasionally. With variable renewables on the grid, there is a risk that, even with backup, there may be shortfalls and supply perturbations. Given the integrated system, and with average annual capacity factors for renewables rising (for example, to over 60% for offshore wind) as the technology improves, the probability of overall system power gaps/shortfalls should be very low, but it is not zero.

However, it can be lowered, and possibly without incurring too much extra cost. A study (of China) has suggested that, as more wind and PV capacity is built and spread around the country, the percentage of it that could be relied on, statistically, as 100% 'firm' capacity will rise, so reducing the need for backup and cutting the cost of getting to higher reliability (Hu et al. 2019). So system reliability may improve as the system expands.

Some are also optimistic about the role that electric vehicles can play in the new power system, given that their batteries could provide a vast distributed energy storage system. That is debatable. While vehicle-to-grid (V2G) exchanges from the batteries of electric vehicles (EVs) might play a role in balancing, as I mentioned earlier in this chapter, it is also the case that the wide-scale adoption of EVs would expand demand for power significantly, maybe by nearly 10% (Evans 2017); others have put it higher. That extra demand would also be variable, leading to further balancing problems, which some say could be severe (Wyman 2019). For example, demand for EV charging would be likely to peak in the evening at the same time as heating demand peaks, at least in the winter. So, if heating is to be supplied at least partially from electric heat pumps, it may be necessary to impose blocks on EV charging at peak demand times in winter, an extension of the demand management idea.

However, even with that system in place, there could still be EV power demand peaks, whereas in terms of energy system management issues most of the other transport options should involve fewer balancing and energy supply problems. Although there are mobility demand peaks, public road or rail transport is much less energy intensive per passenger mile. So too are walking and cycling!

Transport needs clearly heighten almost all the problems and issues facing us as we seek to reduce emissions, and I will be looking at this some more in the next chapter in the context of wider choices about the way society and the economy might develop. While technology can help to some extent, in general the transport issues, and many of the technical/balancing problems I have discussed so far, would be easier to solve if we adopted new patterns of energy use, more energy-conscious lifestyles and consumer behaviour. So in the following chapters, moving beyond just the technology, I turn to look at some of the larger issues concerning social and economic change as we try to move towards an environmentally sustainable low- or even zero-carbon energy future.

5
The Limits to a Sustainable Future

Renewables are expanding, but so is energy demand. If demand for energy cannot be cut by technical means and/ or by social and behavioural adjustments, and we want to avoid the full impact of climate change, what are the options? Must global economic growth be halted? Or can (and should) renewables expand to sustain growth?

Sustainable growth?

Technology can help us limit some of our environmental problems, but it is unlikely to be enough on its own. We may also need social change. This chapter explores the limits to technology and then the options for social and economic change. Some say the changes needed will be quite radical and may include an end to economic growth. Others, however, say that sustainable growth is possible with the help of renewables. Is that credible?

As has been indicated in previous chapters, the potential of renewables at various scales is very large, sufficient in theory to meet all global energy needs into the far future, if the necessary technologies are fully developed and demand is managed effectively. Although there may be other limits to economic growth, some say that renewables will allow, and indeed enable, continued growth both in the economy and in energy use by removing the limits imposed by climate change. So it is claimed, *if* economic growth is what we want, then in theory, as far as energy supply and its impacts are concerned, renewable energy should be able to deliver it, and energy issues will not be a constraint on economic growth.

Whether that is actually the case is debatable, and I will be coming back to that question below but, before I do, there is a wider issue: *are* we actually still aiming for continued economic growth? Is it actually possible? Some say that, on a planet with finite resources and carrying capacities, there are other social and environmental limits to growth, in addition to climate change, and so they look to a more sustainable stable-state economy (Fridley and Heinberg 2018).

The debate on growth has become rather polarized. Bloomberg's Michael Liebreich has put the essentially technological/market-fix view that economic growth is vital for humanity and that it is possible to continue with it (Liebreich 2018). Putting an ecological view, Professor Tim Jackson from the University of Surrey says that it is not, and it is lethal for the planet (Jackson 2018).

The growth debate certainly opens up some key issues, for example of non-energy-related resource usage, crucially, materials and water. While growth clearly has environmental impacts, there are also issues in relation to social equity: without growth, what of those who currently live at or below subsistence level? Growth may be their only hope. Some also see growth as vital to sustaining and expanding employment opportunities. That will certainly be an important issue for the proposed energy transition. Although old jobs will be replaced by new jobs, the changeover will need careful handling as part of a 'just transition' process. Simply reciting the 'maximalist' eco-view that 'there will be no jobs on a dead planet' may not be enough to deflect concerns about employment security (Bairstow 2019c).

This chapter explores some of these issues, asking whether there are limits to growth both in terms of energy use and in wider social, environmental and economic terms. It then moves on to ask what can be done if these limits are real – what are the social change options?

As a starting point, though, perhaps the first question to ask is: are there limits to *technology*? The use of renewables in particular is, as noted above, sometimes seen as a way to avoid wider social and environmental limits to energy use and more generally to economic growth.

The limits to technology

The International Energy Agency, amongst others, has warned that despite the rapid expansion of renewables we will be unable to meet climate-safe emissions-reduction targets since the use of fossil energy is still expanding (Deign 2018). As Box 3.1 indicates, that is how some oil company scenarios depict the likely future situation, and certainly so far the positive climate impacts from the growth of renewables have been blunted by the continued expansion of fossil fuel burning. Basically, one inference from this is that renewables cannot expand fast enough to make much of a difference.

This issue can be looked at in a number of ways. In one version, the problem is growth: we are using more and more energy, doing more things with it and needing more, with renewables being unable to meet that need and fossil fuels filling the gap. Put another way, since overall energy demand is rising, although, as renewables expand, the fossil *share* may be less than it would otherwise have been, the *scale* of fossil fuel use may still grow. Either way, renewables are not winning, and some say they never can since they cannot expand sufficiently to keep up with demand.

Is that really the case? And are renewables fatally limited? REN21's 2018 Renewable Energy status report noted that the share of modern renewables in final energy consumption had grown on average by around 5.4% over the past ten years, whereas global energy demand had only grown by 1.7% (REN21 2018). So, far from falling behind demand growth, they were running well ahead, although not enough in absolute terms to avoid a rise in emissions.

The adoption of more stringent emissions controls, climate taxes and other climate policies might speed up the phase-out of fossil fuels, but can renewables expand fast enough to fill the gap? There are constraints. I looked at some of the *specific* reasons why renewable expansion might be constrained in chapter 3 in terms of objections to individual technology options. However, there may also be more general, generic constraints on what renewables can achieve, most obviously in terms of the total amount of energy available from them.

The first point to make in this context is that the theoretical resource is vast, probably more than we could need, even given unlimited economic growth into the far future. For example, it has been noted that 'within six hours, deserts receive more energy from the sun than humankind uses in a year' (Desertec 2019), while the total wind resource, including offshore and in the upper atmosphere, has been put at about a hundred times more than the total current global primary power demand (Marvel, Kravitz and Caldeira 2013). There are also very large biomass, geothermal, hydro, wave and tidal resources in some places around the world.

However, it will not be possible in practice to exploit all of these resources, given that there are technical, locational, economic and environmental constraints. For example, the value of biomass is fundamentally limited as an energy source by the very low efficiency of plant photosynthesis, and there are land-use constraints on its expansion as an energy source. There are still land-use issues, but, fortunately, solar PV energy-conversion efficiency is much higher, as is wind energy conversion, and we are doing well with these and other new technologies because costs are falling. We are also learning how to use these resources more efficiently. For example, the DNV GL consultancy's Energy Transitions Outlook says that 'electrification and its inherent efficiency will contribute to humanity's energy demand declining from the mid-2030s onwards. Global expenditure on energy, as a percentage of GDP, will fall 44% by 2050' (Lovegrove 2018). I will be looking at that argument later, but essentially DNV GL is claiming that the need for primary energy can be reduced.

Optimists also argue that energy-saving innovations will allow us to continue to cut demand (Lovins 2018), with some suggesting that it will be possible to reduce global energy demand to 40% *lower than that of today*, despite rises in population, income and activity (Grubler et al.

2018). Jacobson's new scenario has global energy demand actually falling by 57.9% by 2050, due to fuel substitution and energy-efficiency upgrades, and with renewables then meeting the residual demand in all sectors (Jacobson 2019a).

While Jacobson, and also LUT/EWG, claim that we can get to 100% by 2050, others are more cautious, especially in relation to the heat and transport sectors. IRENA claims that we can get to 86% global *power* by 2050, but that still leaves some way to go and the projections for heat and transport, even those of some optimists, are often lower (IRENA 2018, 2019c).

Certainly, pessimists are concerned about possible technical problems and potential limitations in all sectors and end uses. The transport sector is particularly hard to deal with. There are technical fixes but they may have short-comings. For example, electric vehicles are usually seen as cutting emissions compared to fossil-fuelled cars, which will be the case in direct emissions terms if they use renewable electricity for charging. However, the energy used/embedded in EV construction and in making their batteries also has to be taken into account. When that is done, the comparative lifetime carbon-saving figures, although still positive, do not always look quite so good, depending on where in the world this energy is sourced (Hausfather 2019).

There are also some non-energy environmental issues with EVs, which may undermine the value of these savings (see Box 5.1).

Box 5.1 EV impacts and alternatives

Electric vehicles may reduce emissions, but they make no contribution to reducing road congestion or the need for more roads and more parking spaces. Moreover, EVs have rubber tyres and, as with other rubber-wheeled vehicles, running on asphalt roads can give rise to dangerous micro-particulate air pollution (Harrison 2017). It may be that this will be as important as avoiding emissions from fossil-fuel engine car exhaust pipes, especially because, due to their battery weight, EVs tend at present to be heavier than conventional cars. That may also apply to hydrogen-powered cars, including those using fuel cells, although they can be lighter, especially if hydrogen can be stored as chemi-absorbed metal hydrides, rather than as a liquid in heavy cryogenic tanks. So it may be a materials science/technology issue. However, there are materials issues and a need for new options. For example, as I will be discussing below (Box 5.2), there may be resource limitations and eco-impacts associated with material extraction for some of the materials currently used in the batteries and motors of EVs (NHM 2019).

Technology may improve for both the EV/battery and hydrogen car options, with on-board solar PV battery charging also being an EV possibility (Mace 2019), but it seems fairly clear that, if we want to reduce energy use *and* environmental impacts, then for most transport needs we will need to move away from private vehicles to public transport, which is much more efficient, with electric- and hydrogen-powered buses, trains and trams reducing emissions even more. Biogas-powered heavy vehicles may also be an important option for some uses. However, it has been argued that we need to reduce overall demand for powered personal mobility while enhancing public transport, walking and cycling and perhaps even travelling less: technology can only help to some extent (Vaughan 2019). That is not to say it is irrelevant. A UK Energy Research Centre study suggests that lifestyle changes, coupled with a big EV push, could in theory cut UK transport emissions by 57% by 2050 (Brand, Anable and Morton 2019).

The message seems to be that, although there are technical options, we may also have to change our mobility modes, habits and expectations. The big transport issue is of course air travel, where the social change and technical limit issues come to the fore strongly. To be fair, although it usually attracts much environmental ire, aviation's share of energy use, and the emissions from it, is relatively small at present. Commercial aviation only burns 2–3% of the world's fossil fuels and, although estimates vary, an IPCC assertion was that aircraft cause about 3.5% of all global warming attributable to mankind. In 2016, aviation was said to account for 3.6% of the total EU28 greenhouse gas emissions (EASA 2019).

However, demand and emissions are growing. In Europe, aviation emissions have doubled since 1990, and globally they could, without action, double again or even treble by 2050. So one study claimed that, unless checked, aviation emissions may account for up to 25% of the global carbon budget by 2050 (T&E 2018). Are there any solutions?

As the IEA has said, continued efficiency and operational savings are possible and there are substitute fuels emerging, including biofuels and synfuels like hydrogen, as well as electric propulsion, using green power for pre-flight on-board battery charging (IEA 2019d). There are certainly technical options (Bowler 2019; Ellsmoor 2019a) but, while the cost of short-haul flights, based on untaxed fuel, is so low, demand (and emissions) will be hard to tame. Moreover, left unchecked, demand growth could soak up renewable resources, green hydrogen and so on, which were, arguably, better used for other purposes, and also accelerate demand for biofuels, with potentially large environmental problems (T&E 2018). About the best that can be hoped for, in terms of technical fixes, is that solar-powered battery-assisted flight will prove viable beyond the small-scale lightweight systems so far developed.

Wider energy and resource issues

Beyond specific sector issues and problems like this concerning the limits of technology and the use of renewables to reduce emissions in some areas of energy use, there are also wider, more generic energy and carbon debt issues related to the use of renewables. For example, it is sometimes argued by pessimists that renewable energy technologies cannot be a major help in dealing with climate change since, they claim, too much energy is needed to make/build the hardware, compared with what the systems can produce once built. More specifically, it is argued that the 'energy return on energy invested' (EROEI) ratio is low for most renewable energy systems, compared to the historic EROEI for fossil fuels. So it is claimed we will be seriously handicapped in making a changeover: we will need more input energy to produce the same output energy. Thus a group of Finnish academics have argued that 'because economies are for the first time in human history shifting to energy sources that are less energy efficient, production of usable energy (exergy) will require more, not less, effort on the part of societies to power both basic and non-basic human activities' (Järvensivu et al. 2018).

There is something of a paradox here in that, whatever happens (even given a massive cutback in energy demand), we will have to replace our power systems as old plants become obsolete, so there will be a carbon debt for that too even if no renewables are used, and it will be made worse by the falling EROEI of fossil fuel, if that is what the replacement plants use. That said, it is true that the construction/manufacture of renewable systems will need significant energy inputs and that the EROEI for some renewables can be low. That was the case for PV solar in its early days: it initially involved very energy-intensive cell production. However, that has now improved, with EROEI energy out/energy in ratios of 20:1 or more being common, and with energy payback times of months not years. Wind-turbine EROEI has also improved, with good systems getting up to 80:1. That may be a high estimate but, for comparison, although it cites lower renewable EROEI, a recent study has claimed that the EROEI for fossil-fuel end use is now running at around 6:1, and potentially as low as 3:1 in the case of electricity generation and use (Holder 2019).

Nevertheless, although most renewable-generation EROEI may not be too bad, it is true that balancing systems will be required for variable renewable inputs, and some of these systems (pumped hydro apart) may have poor EROEI. It is a complex debate, diverging over the methodology and the value of comparisons (Raugei 2013; Weißbach et al. 2013, 2014) and over the resultant numbers (Hall, Lambet and Balog 2014). For example, some EROEI studies just measure the energy used in materials and construction (the so-called 'embodied energy') and do not include the energy needed to produce the fuel needed to *run* (some) plants, which can be a significant factor for the highly processed fuel for nuclear plants. Renewables like wind and PV do not need fuel to run, so is it fair to make comparisons on that basis? Otherwise, with the fuel production debt included, nuclear-plant EROEI is typically low, at around 15:1, and that is likely to fall (to maybe 3:1 or less) as high-grade uranium reserves get scarcer and harder to access and process. That makes nuclear an even less likely contender as a replacement option for fossil fuel.

It is also hard to know what time frame to use. Hydro plants can last for hundreds of years without needing much upgrading so, despite their large energy/carbon investment in concrete, they can have lifetime EROEI ratios of 200:1 or more. Is it fair to compare them with devices which only last for 20 or 30 years? It is true that, historically, the EROEI for fossil energy was high but, as indicated above, it is now falling: we have to accept a new future, and for most renewables the EROEI is reasonable and should improve as the technology develops. For example, it has been claimed that airborne wind systems could have an EROEI of 100:1 or more. Airborne wind may be a non-starter, but there may well be other clever ideas to come. Is it sensible just to stick with current technology when making comparison with earlier EROEI?

A similar and linked resource debate concerns the use of scarce *materials*. Most technologies make use of some scarce materials, and it is not clear if renewable energy systems make especially heavy use of them, or of less scarce but energy-intensive (and therefore carbon) materials like steel, aluminium, copper and concrete. However, whatever they are used for, there are social and environmental impacts associated with producing these materials and with the so-called 'rare earth' materials in particular. That includes neodymium, which is used in power-generator magnets. So there may be serious issues to face if we have to use these materials for any new energy system, including renewable energy and storage systems, the latter being an especially rapid growth area for materials use (Colagrossi 2019; Courvoisier 2019; H. Liu 2016; and see Box 5.2, which tries to put the issues into perspective).

Box 5.2 Material shortages

A full life-cycle resource analysis has suggested that renewables could supply the world's entire electricity needs by mid-century without major problems with resource (materials) use or associated eco-impacts. It assessed the whole-life costs and associated materials-sourcing impacts of solar, wind and hydro technology in relation to the demand for aluminium, copper, nickel and steel, metallurgical grade silicon, flat glass, zinc and clinker. The overall conclusion was that 'bulk material require-ments appear manageable but not negligible compared with the current production rates for these materials. Copper is the only material covered in our analysis for which supply may be a concern' (Hertwich et al. 2014). Issues still remain for concrete, although there are lower-carbon versions emerging (Nazari and Sanjayan 2017).

The use of rare earth minerals and lithium (for batteries) is also growing (Arrobas et al. 2017). That may lead to problems both of impact (mining and processing these materials are polluting and can be hazardous) and also of access. Most of these materials are not widely available globally. However, post-first-use recycling of some of these materials can help, and in some cases substitutes can be found or systems redesigned to avoid or limit the need for them. It has also been argued that more efficient use of the scarce materials can help (evidently, there is some wastage during production). The discovery of new reserves in a wider range of countries, and possibly also via seabed mining, should also help to widen access (Than 2018). I will be looking at the geopolitical implications of these resource access issues in the next chapter. But wherever it is carried out, the environmental impact of extracting these materials may be serious, especially in terms of lithium for batteries for EVs, as well as phones and laptop PCs (Katwala 2018; Lombrana 2019). We may need to develop alternative approaches to meeting our needs, including for energy and energy storage, to become less reliant on some of these materials.

Access to *water* resources could also be a major issue for some energy technologies, made worse by climate change. The water supply issue is usually more of a problem for fossil and nuclear plants, which need large amounts of water for cooling, but so do some renewable energy technologies, notably concentrated solar power (CSP) plants. CSP is best located in areas with direct, non-diffuse solar, such as deserts, and water is one thing such areas do not have. Air cooling is an alternative option for CSP but is less efficient. More likely, water will have to be piped in, which suggests that locations reasonably near the sea would be best, as in North Africa. Water is also needed for cleaning CSP mirrors and lenses, and for PV solar arrays, small or large. That will mount up in the years ahead, although there are self-cleansing designs and also electrostatic technologies for dust removal, as pioneered for the PV cells on Mars/lunar rovers. Less easy to deal with is that water is vital for biomass growing, and hydro plants also obviously need water. As already noted, with decreased rainfall in some areas in recent years, biomass growth is being constrained and the output from some hydro plants has fallen (Baños Ruiz 2018; Gabbatiss 2018). Water scarcity issues may thus add to the problems of both hydro and biomass.

Technical fixes and the pace of technical change

As can be seen, renewables face a range of technical and resource problems, although there may be solutions to some of them. Some pessimists tend to view attempts to ameliorate or limit problems as mere 'technical fixes', which will not deal with what they see as a fundamental weakness in the case of renewables. While avoiding the direct use of fossil and fissile energy sources, these fixes still inevitably rely on the use of other materials and resources, and on energy to make the materials needed for their construction, as well in some cases as water and land.

Whether that all adds up to a damning case against renewables arguably depends to some extent on how much and how *rapidly* they are expanded. Some worry that, in addition to the materials issue discussed above, rapid expansion may be constrained by increasing scarcity of fossil energy sources. For example, one environmentalist has argued that renewables 'currently require fossil fuels for their construction and deployment, so in effect they are functioning as a parasite on the back of the older energy infrastructure. The question is, can they survive the death of their host?' (Heinberg 2015).

As noted above, while the fossil resource is still substantial, it is true that the EROEI for fossil fuels is falling and also that we want to phase out use of the latter. So it could be argued that we should be reserving as much as possible of whatever fossil energy is left to support the process of building up renewable-based systems, rather than just burning it off for no long-term gain. The same might be said for rare earth materials. We should not squander them to meet less urgent needs, although obviously that has to be put in perspective; we cannot freeze development in all sectors, privileging just renewables. For example, there are important non-energy uses for fossil and other resources so some should be reserved for those too, although it may be that we will want to produce fewer plastics (from oil), given their ecosystem impacts.

Resource scarcity is not the only issue. Using fossil fuels in the interim to build the renewables system will also lead to emissions debts and eco-impacts. We will want to limit those as much as possible. However, to some extent, the first wave of renewables can provide energy to support the construction of the next wave in a self-sustaining and expanding 'breeding' process, avoiding fossil fuel emissions, with renewables gradually taking over from the use of fossil fuels for materials production. That may take time, but the 'energy in/out' returns of most renewables are improving, so as green energy from these better technologies is increasingly used to build the next wave of technologies and produce materials for them, the net fossil fuel debt will fall.

In emissions terms, that will be doubly important, given the falling EROEI of the fossil resources and similar falling EROEI problems with fissile fuel, if any of that is used. CCS, BECCS and/or other negative carbon options might help compensate for any fossil emissions during the initial phase of the transition process, especially if we are aiming for a rapid start to the changeover, until new renewables can take over, though arguably reforestation might be a better bet for that compensation.

So there may be ways to limit some of the resource problems, depending in part on how quickly we want to make the transition and the success of the various technical fixes. However, a relatively slow pace of overall change, allowing time for renewables to take over more of the materials supply/construction load, as well as energy supply, may not be enough to avoid major climate impacts. If we need to go faster and further, what are the options?

Optimists will say 'more renewables', deployed faster, coupled with more strenuous energy saving, and with any construction emissions offset by using negative emissions technologies. Pessimists will say that will not be enough and that growth in energy demand will continue to overwhelm the carbon savings. One grand technical-fix solution to this conundrum often offered is to go for nuclear and/or retain fossil fuel use, but with CCS added to make it nearly carbon neutral or, if all else fails, suck CO_2 from the air (N. Johnson 2018; Myhrvold 2018).

As I have indicated, there are many problems with these grand technical fixes, arguably more than for the renewables technical fix. This is not the place to reprise all the technical, economic, social, safety, security and environmental problems associated with the use of nuclear fuels, but their finite nature and energy-intense production (with low and falling EROEI) suggest that, even if we could accept the operational and economic problems of using this source, relying on it is not a long-term option. Similarly for fossil CCS, as I argued in chapter 4. There may be sufficient fossil reserves for many decades ahead, even a century or so, depending on use rates, but can we really keep burying the results underground or suck CO_2 from the air, in effect, to make room for more? At best, air capture might be seen as a remedial option, cleaning up past emissions, but more likely it will be used to make new fuel, in carbon capture and utilization (CCU) mode, the problem being that, when burnt, these synfuels would just add emissions back again (Elliott 2019d).

Some look to a hybrid approach: a bit of renewables, a bit of nuclear and a bit of CCS or CCU. Indeed, that is what is assumed in some of the more conservative scenarios I have looked at, and there have been renewed calls to keep nuclear in the mix (IEA 2019e). However, as I have indicated, there are problems with this compromise. For example, inflexible nuclear and variable renewables do not work well together on the same grid (see Box 4.4), and, given the high current cost of nuclear and CCS, there is a risk that an ostensibly 'mixed' system would end up with most of the financial resources having to go to them. Therefore renewable development would be slowed. A recent report for EU warned that 'in the long term, it is possible that policy tools may be necessary to ensure that CCU does not perpetuate the use of fossil fuels' (Searle and Pavlenko 2019), and there are certainly issues with combining carbon removal with zero-carbon generation in 'net zero carbon' policies (McLaren 2019).

Arguably, we need to make a decision as to which way to go or else risk not doing anything well. Some argue that a mixed system would ensure diversity, with several eggs in several baskets. That idea clearly has merit, but renewables represent a wide range of very different options, some new, some old, whereas nuclear fission is basically one already well-tried option, while CCS/CCU is a single, as yet mostly untried, option, both of these energy options having no long-term future. If we want diversity, to spread risks, then renewables can provide it and they offer expanding opportunities into the long term.

However, if renewables used alone are deemed not to be sufficient, for whatever reason, and if the mixed system is also seen as unviable, that leaves us with one more set of theoretically possible technical fixes: planetary geo-engineering. This includes ideas with uncertain and perhaps irreversible environmental impacts, for example trying to increase greenhouse gas absorption by seeding the seas with chemicals (Weston 2019), and also for reducing solar gain by injecting light-blocking aerosols into the upper atmosphere (Carrington 2018). There have even been proposals for giant sun-reflecting arrays to be put in geostationary orbit around the earth to reduce solar input.

While some of the less wild ideas may be worth investigating, in general it would seem to make more sense to use the solar energy than to try to block it out, just to allow us to continue to burn fossil fuels. Geo-engineering seems, in most cases, to be a desperate and potentially dangerous last-ditch technical fix, and there has been strong opposition to it, along with the various carbon removal options (Muffet 2019).

However, if these options, along with all the other technical fixes looked at above, including renewables and energy saving, are seen as unviable or undesirable on the scale needed, and if we want to deal with climate change, then we are left facing head on the need for more radical social and economic responses, including perhaps the end to growth and the economic and social system it supports and relies on. That is no longer a fringe issue. Leading US journal *Foreign Policy* has asked whether economic growth can continue, even so-called 'green growth', implying possibly massive social and economic change (Hickel 2018).

Social change

There may be many reasons why we might want radical social and economic change, but in terms of reducing energy use the social change options can be viewed in various ways. At minimum, it means seeking change in consumer behaviour, so that energy is used more efficiently and, hopefully, the overall use of it is reduced. Some of that will involve adoption of new, more efficient technology for energy use, or perhaps domestic-level energy generation. Beyond that come actual changes in consumption patterns, perhaps aided by time-of-use pricing systems. As I have indicated, that may be able to shift peaks and thus cut the need for peaking plants, but it may not cut overall energy use significantly, just re-phase it. To go further means cutting actual energy use, either voluntarily in response to appeals to 'go green', or involuntarily in response to punitive pricing or even direct rationing of energy use. Both options are likely to be very invasive and could be socially divisive and potentially regressive, depending on how they are managed (see Box 5.3). You can see why *technical* fixes, even radical ones, have their attractions as seemingly less draconian options.

Box 5.3 Energy/carbon rationing

Some look to the imposition of constraints on energy consumption, via personal energy or carbon credit accounts. Consumers would be allocated energy or carbon credits to limit their overall consumption of energy and fuel. These credits would be tradable, so that consumers who managed to use less could sell any excess to those who were less frugal, creating a market for credits. The annual allocations for everyone would be gradually reduced.

Enthusiasts say, not unreasonably, that personal rationing schemes would have an immense educational value, making people very aware of their energy use and/or carbon debts, leading hopefully to behavioural change. But some people may see personal carbon rationing as an unwarranted imposition, for example requiring invasive policing, given the potential for evasion and abuse.

In the worst case, there could actually be a net increase in emissions. The rich and energy-profligate could simply buy credits from the poor to avoid the overall cap limits while, given their high value to the rich, the poor would be tempted to sell their credits and try to buy in dirty 'off list' energy as a replacement. Of course, in the best case, at least for the climate, given the increasing cost of energy and credits, the rich might cut back to some extent and/or invest in efficiency/self-generation, while the poor might decide to do without some energy services in order to continue to sell off their credits. So in that case there might be some reduction in emissions. But at what social cost? In the end, it is not that different from the rationing effect of imposing higher prices on energy.

While technical fixes may have their attractions, they also have social impacts, due for example to the extra costs that some say will be involved with moving to green energy. As mentioned in chapter 1, fears about the rising costs of support schemes for renewables have led to warnings from otherwise divergent camps about energy justice and fuel poverty (Beisner 2019; Monyei et al. 2018).

How significant is this problem? Will prices really have to rise? While, as was noted in chapter 1, green power generation costs have been falling, it may still be the case that, as noted in Box 2.2, retail prices may not always fall. Indeed, they might rise, at least initially, as the green energy programme expands, although it is also argued that those costs may be offset by energy savings.

In the UK context, looking at the various green energy subsidy schemes and tariffs, the government's advisory Climate Change Committee has claimed that 'for households, the average costs so far, of £105 per household per year in 2016, have been more than outweighed by savings from improved energy efficiency: energy bills fell £115 in real terms from 2008 to 2016. That balance will continue to 2030.' It added that, subsequently, 'our scenarios involve an annual resource cost of around £4 billion in 2050,' down from £7 billion currently, and around £12 billion by 2030, and it claimed that 'overall bills need not rise as a result of climate policy' (CCC 2019).

There may be a degree of wishful thinking involved here in that, in practice, it has proved hard to get energy-saving projects running successfully: some have been abandoned or withdrawn in the United Kingdom (Elliott 2019d) and, globally, progress has been slow. However, that does not mean that it is impossible and, longer term, as savings build up and the cost of using fossil fuel is avoided, overall costs should stabilize and maybe even fall.

In chapter 1, I quoted the European Commission's claim that, under its proposed renewables-led transition, 'by 2050, households would spend 5.6% of income on energy-related expenses, i.e. nearly 2 percentage points lower than in 2015 and lower than the share in 2005' (EC 2018).

Nevertheless, there may be a transitional period when costs may rise, especially if we want to move *faster* than currently planned and also tackle the harder green heat and transport issues more effectively. In that case, it will be vital to ensure that any extra costs that are incurred are shared fairly so as to avoid conflict and resistance. There are many options. It could be done equally, per capita, or pro rata for energy uses across all consumer categories. Or should safeguards be built in for the less well off? And should high-energy users be hit harder to get them to change? So there are many choices and issues, a crucial one being who would consumers and the public trust to make such decisions and then manage the change, given the arguably very poor past performance of governments and power companies in relation to energy price management (Becker et al. 2019)?

In terms of a socially equitable transition, the CEO of the UK Climate Change Committee has said:

> while the economic costs of decarbonisation overall may be smaller than we thought – potentially allowing the UK to go further for the same cost envelope, I doubt we will make further progress without a thorough review of how these costs are distributed – and the appropriate strategic policy levers. It follows that we must consider the appropriate balance of 1) cost for the Exchequer; 2) costs on

the consumer; and 3) economy-wide costs. And we must make use of the right tools – carbon pricing, tax, financial incentives, information or regulation. (Stark 2019a)

Wider education on the need for change can help, but there can be a somewhat moralistic or paternalistic view underlying some of the proposals made for reducing energy use and emissions, which can be counterproductive or at least provocative. However, equally, given the dangers of climate change, tough measures might be seen as a matter of long-term survival. While that may be true, it will not be easy to win agreement to changes, given that views on climate change and policies to deal with it can vary, leading to political and cultural polarization (Martin 2019).

Clearly, there are some serious societal issues and also disagreements. Some say we are wasting much of the energy we produce with inefficient systems, and also arguably using it for frivolous or even unethical consumption, given that many people worldwide are not able to enjoy such luxuries. On that view, we should improve efficiency, cut excessive energy use and also, in some of the more radical variants, find ways to redistribute wealth and power so that everyone can share in the benefits of technology, with growth being stabilized. In most prescriptions, renewables are included as part of the solution since they are seen as being cheaper and more localized, so potentially offering more opportunities for local control and ownership of energy systems. It seems to be implicit in this sort of idealistic view that they can deliver a sustainable future at reasonable cost, given lower growth expectations.

Critics sometimes claim that this is not the case: they insist that renewables will be more expensive and less able to support a high standard of living, much less growth, which, it is claimed, is what is needed to allow those living at subsistence level to move ahead. Alternatively, some welcome this constraint and say that it means we all need to reduce our use of energy. For example, Australian academic Ted Trainer has argued, in a critique of Delucchi and Jacobson's '100% renewables' paper, that renewable energy could supply the world only if the world 'embraces frugal lifestyles, small and highly self-sufficient local economies, and participatory and co-operative ways in an overall economy that is not driven by growth or market forces' (Trainer 2012). In response, Delucchi and Jacobson say that 'This vision may or may not be desirable, but it was found in our study not to be necessary in order to power the world economically with wind, water, and solar energy' (Delucchi and Jacobson 2012).

Trainer clearly thinks that renewables cannot support growth. However, he may also risk undermining them. Indeed, it is almost as if he does not want them to work, so that we have to get on with what he sees as the more important social changes. Certainly, some see radical social and lifestyle changes as vital as part of an urgent political and economic process of change (S. Johnson 2018).

Does that also mean an end to economic growth, as Trainer clearly thinks? There have been powerful cases made against it, pointing to the social and environmental problems it has created and arguing that prosperity and economic growth are not necessarily the same thing (Jackson 2009). Others have argued that energy use and economic growth can be *decoupled* by using more efficient technology, getting more from less. However, as I have discussed above, there may be limits to how far that is possible. Then again, some say these limits can be transcended if we make radical changes in society and economy. Moreover, far from reducing the quality of life, a stable-state economy, freed from the need for relentless economic growth, could attend to social needs in a more equitable way. The US Post Carbon Institute has been struggling with these issues for some time, looking to stable-state economics (Fridley and Heinberg 2018; Heinberg 2011).

This is not the place to map out blueprints for that or any other social future, but it does seem clear, first, that the current socio-economic model is in trouble, and second, that whatever might replace it has to move towards what has been called *sustainable consumption*. There is a long way to go on that, with many affluent people still wedded to conspicuous consumption as part of their lifestyle and many poorer people wanting to be able to adopt that. There are exceptions, with local community 'voluntary simplicity' and low-consumption initiatives spreading in some affluent countries, usually on the basis of commitment to green ideals. However, although often inspiring, they may remain marginal. As one study put it, 'grassroot actors may be more progressive in their demands and actions, but their overall impact is most likely limited' (Boucher and Heinonen 2019).

Arguably, what is needed is wider, more systemic, change. Looking broadly at wider socio-technical change patterns, some argue that the process of moving away from material consumption is actually already underway in that we are moving to a post-capitalist 'dematerialized' world in which the economy is based increasingly on trade in information, which of itself has no energy or material content (Bastani 2019). Put baldly like that, this seem overly naive in that it takes energy to shift and process information and to make the equipment to do that. For example, it has been claimed that, globally, about as much energy is now being used by information and communication technology (ICT), including mobile devices and servers, as by air transport (BBC 2018).

It may be possible to reduce ICT-related energy use (even if blockchain use spreads), but it is plain that there is more to life, and the economy, than ICT! People still need food, shelter, heat and mobility and myriad other services. Energy use can be reduced in each of these areas, and, hopefully, renewables can be used to meet most of the residual needs, but perhaps not all of them if growth and ever-expanding consumption continue to be the mainstay of the economy. There are limits, not all of them energy-resource limits, and decoupling may not be possible beyond a certain level.

Ultimate limits

There may be some constraints on economic growth that cannot be relieved by the rapid expansion of renewables. Although the analysis in this book suggests that renewables may allow us to expand energy use while avoiding climate limits, I have also noted that there may be non-energy/ non-climate-related constraints on economic growth. There are limits to other resources beyond energy, as well as social constraints, concerning acceptable lifestyles. It is hard to see exactly what may happen in the future in terms of those wider issues, or even in terms of energy, but it is likely to depend to some extent on what sort of society we are talking about.

Some look to a future in which automation improves production efficiency, so that energy and resource use falls and so too does the drudgery of work. There are problems with that. On past experience, technology may eliminate some boring and unpleasant work but not all, and a future without some form of gainful employment, however defined, is not attractive to all (Elliott forthcoming).

Certainly, it is not clear that a fully automated high-technology future will be welcomed by everyone or in fact deal with our environmental problems. We will still need resources, even in the most technically efficient system. It may be worth taking some extreme examples to see how that might all play out.

In one extreme *ecomodernist* vision of the future, nuclear power (maybe fusion) provides the bulk of energy and genetically modified syn-food is grown hydroponically/in vertical farms, all based in vast new cities, with the rest of the planet left to run wild (Brook et al. 2015). It is, in effect, a giant autonomous 'space colony' or spaceship on earth! No doubt we can argue the details: the cities may have to import specific materials. They are also likely to want to trade with other such cities to maintain their economies. Indeed, they may want to expand their economies, maybe setting up offshoot colonies. So growth could be important and automation may be widespread, with social and economic hierarchies also likely to exist. That may not all be inevitable. The city-states might be run democratically. However, for the sake of argument and details aside, this does represent an extreme vision of what might happen.

At the other extreme is the fully decentralized renewable energy-based network of smaller stable-state communities distributed across the planet, based on strong community participation and cooperatively run and owned enterprises. The technical emphasis would be on local power generation, on local organic farming and on artisanship and craft trades. There might be some trade between the communities, but no drive to expand and, presumably, no cities (Bradford 2019; Trainer 1995). Again, we can argue the details. For example, there might be a need for more trade in order to meet all their needs, including for energy and energy balancing, although a 'fair trading' stance is likely to be taken. However, that and the other generally progressive social and political aspects may not be inevitable. We have had societies in the past something like this, for example in feudal times, and they were not exactly egalitarian.

How do these admittedly very hypothetical and caricatured alternatives stand up in terms of sustainability? The centralized hyper-cities are totally dependent on the success of advanced technology, so whether they can be sustainable or not depends on that success and possibly on their growth to sustain the system. The alternative decentralized communities are also dependent on technology but arguably to a lesser extent, with less need to grow. Life might be more frugal but arguably more wholesome, and the environmental impacts ought to be low.

However, one 'elephant in the room', possibly in both cases, is population. In the past, population and growth have been intertwined. Summarizing grossly, when material growth became possible (through technology, trade or conquest), population boomed, but growing population (or sometimes falling population) also at times stimulated economic growth. Nowadays, we tend to see population as a threat: more people, more consumption and more environmental impact. The reality is, arguably, more complex. The small minority of rich people, mostly in affluent countries, tend to consume far more than the larger number of poor. That is socially divisive, as well as environmentally problematic, but it could get worse, given that, reasonably enough, most poor people want to become rich. If that happens, and if population also continues to expand, some fear that energy use in absolute terms and per capita will rise and overwhelm the emissions savings from decarbonization measures. In addition, population growth would also impose other impacts on the ecosystem.

However, this is not inevitable. Technology allows us to control the birth rate and clearly that option should be available to all; and affluence also often leads to decreased birth rates since, arguably, large families are then less needed for economic survival. So some are hopeful that the global population will begin to stabilize if affluence spreads.

Population growth rates do seem to be declining in many countries – indeed, in some they are falling below replacement rates – although in terms of climate impacts that may be too slow and affluence may *not* spread rapidly, or at all. In which case, we may need tougher social policies to encourage smaller families. Not an easy or popular prospect, whatever the type of society, though it might be easier in the decentralized version sketched out above, with the local impacts of population pressures possibly being clearer. Then again, that might also be said of the densely populated hyper-city version.

Some say that big, seemingly intractable issues like this will be resolved, and solutions imposed, as a consequence of resource limitations, as has happened in the past in response to scarcity of food and of productive land for food growing, and of access to fresh water. With soil fertility falling and fresh water being polluted, we may now face similar problems, unless we change our ways. Though it could be that falling human fertility, due to pollution, may resolve the problem at source.

Clearly, quite apart from the gross inequalities apparent around the world, we live in worrying times in which the scale of environmental and ecosystem damage and biodiversity and species loss is becoming ever more apparent (IPBES 2019). Some fear that these impacts may soon be overwhelming, making the specific issue of energy supply relatively unimportant. However, the way in which we source and use energy is arguably a central part of the overall problem and facing up to that must be part of the response to the wider ecological crisis.

Change is possible

The two polar extreme examples of possible futures I introduced above indicate very different types of response, one very centralized and urban, the other very decentralized and rural. In reality, many options exist in between, including ones in which cities and rural areas coexist.

Elsewhere, I have explored some of the energy implications of that (Elliott 2019e). In summary, given their high densities and energy use, many cities are unlikely to be able to generate sufficient green energy to meet their needs from within their own boundaries. For example, a recent EU-based study of urban energy autonomy concluded that 'more than 3,000 European municipalities cannot be autarkic due to insufficient land area' (Tröndle, Pfenninger and Lilliestam 2019). As at present, most cities will have to import energy, as they have always done for food and water. Even with PV solar on all available rooftops, with multistorey multi-occupation high rises, there may not be enough energy to meet demand, and cities will need to import green power from rural areas to compensate. That may lead to conflicts, unless the residents in these areas are paid fairly for the intrusion of wind and solar farms and biomass plantations. So there may be social equity issues to resolve.

More generally, the various possible future socio-technical options may not all deal well with the environmental problems we face. At every point there can be problems, some of them due to human foibles, some due to the shortcomings of technical fixes or market systems. For a small contemporary example of potential problems with a 'high-tech' fix in the transport sector, it is claimed, a little cynically, that with the advent of self-driving vehicles, user mileage may increase (Auffhammer 2019), and that we may even see autonomous cars left driving around instead of seeking a parking space (Kaufman 2019). That is a fairly trivial example and there is no shortage of examples of potentially more substantial problems from other sectors, some of them the result of the way market economics operate. In chapter 2, I mentioned the rebound effect in relation to energy efficiency and also the market cannibalization effect as renewable prices fall.

It may be possible to overcome problems like this, but we do face challenges as we try to move to a more sustainable approach to energy supply and use globally. As I will be exploring in chapter 6, despite all our efforts so far, global emissions remain stubbornly high and are rising, especially in some sectors, notably transport. The World Energy Council at one point estimated that worldwide energy consumption for transportation could grow from 2010 levels by between 30% and 82% by 2050, depending on the pathway taken (WEC 2011). Mobility certainly is a complex and tricky area (Brand, Anable and Morton 2019).

However, being optimistic, there do seem to be some potentially viable social and technical options for the future in most sectors and, arguably, none of the dire outcomes that currently worry us may be *inevitable*. There are things that can be done to avoid them. Some can be dealt with locally, some need global response. Some solutions may be less appealing or less easy than others, but hopefully we can avoid the more panicky, extreme and potentially risky technical fixes and draconian social change responses.

What is unclear is what might drive the necessary changes. There is always resistance to change. Environmental concern is high and growing, but it may not be sufficient to produce change on the scale and pace needed. In chapter 1, I argued that, at least in the energy area, falling costs for renewables might accelerate their take-up rapidly, but that government support and direction might also be needed: it cannot be left just to market pressures.

In the next two chapters, I look at some examples of changes underway around the world, and at their implications, in the energy field. I start first, in the next chapter, with a review of what the necessary changes might mean at the global level in terms of changed geopolitical relations, and how change might be more effectively directed and accelerated locally and globally.

6
The Geopolitics of the Transition

Thinking globally and acting locally is a good credo in that there will be global implications for the changes made, as well as global needs for change. This chapter looks at the geopolitical implications of trying to move to a sustainable energy system.

Global change

The generation and use of energy has global economic, political and social implications, as well as environmental impacts. Dealing with climate change will require global-level responses, and some of the new technologies being adopted, and the process of change that will be needed for them to succeed, will have global social, economic and political implications. This chapter explores the geopolitical issues related to the energy transition.

The global energy situation has been in flux for some time, chiefly in response to battles over the costs of and access to oil, and more recently gas, but energy security and resource concerns have now expanded to include responses to climate change and the development of renewable resources (EUCERS 2018).

Not all of the climate-related responses and associated energy transition plans are global in reach. Some parts of the energy transition may mainly concern national and subnational changes, but even changes that are seemingly just internal to a country can have external implications. For example, many of the older industrial countries are currently shifting their economies from energy-intensive industrial production to service activities of various types. That has helped them reduce their emissions, but many of the industries have moved elsewhere in the world, so the emissions have been exported. In reality, though, they may not have been entirely exported. The *products* that are imported back have an energy content, so-called 'embedded energy' from their production, and arguably the associated emissions should be repatriated.

The 'offshoring' process may be an important short-term trend, and the embedded energy/carbon imports are also a significant issue (Partington 2019a), but as and when the wider global energy transition continues and spreads, and global energy resource-use patterns change, a new global emission pattern and balance (or imbalance) may emerge.

It is interesting in this context that in some of the industrial countries demand for power has reduced, notably in the United Kingdom where it has fallen back to 1994 levels (Evans 2019a). Indeed, overall UK *energy* use has also fallen consistently over the last ten years, despite economic growth continuing (BEIS 2018b). Not all of this is due to the relocation of energy-intensive industry in other countries: one study suggested that the 'offshoring' effect has been broadly similar in scale to that of technical improvements in industrial energy efficiency (Hardt et al. 2018). In addition, some of the reduction has come from the domestic household sector in the United Kingdom (Vaughan 2018) and also in the United States (Davis 2017b).

One of the main drivers for this is that, with prices for energy and for many other commodities and services rising, energy is now being used more efficiently, aided by the advent of new end-use technologies and, in some cases, prosumer self-generation. The end result is that energy utilities are feeling the pinch in the United States (Roberts 2018) and elsewhere. The demand reductions do seem to be spreading. Although energy use and electricity demand globally have still risen, power demand fell in 18 out of 30

IEA member countries over the period 2010–2017, with the IEA saying that over 40% of the reduction was due to energy efficiency within industry, some of the rest being due to the spread of more efficient lighting systems and domestic appliances (Bouckaert and Goodson 2019).

As renewables and energy efficiency are more widely adopted, changes like that may spread across all sectors and around the world, so that, if done right, emissions can fall everywhere. Energy markets will also change. As a result, as and when this process speeds up, a new geopolitics is likely to emerge, with new winners and new losers and new issues to face. They include not just economic competition over resources and markets, but the wider issue of what we are trying to achieve, which must surely be an environmentally sustainable global future.

Views clearly differ on how to achieve that but, while mapping out some of these views, the main thrust of this chapter will be on the opportunities for change, including coverage of the situation in developing countries, especially in relation to equity and justice issues. A sustainable world will not come about by increasing levels of exploitation and inequality.

The process of change also faces other problems and challenges. As was argued in chapter 5, not all will be impossible to deal with and, as this chapter illustrates, some problems are in fact overstated or can be avoided. Nevertheless, there will be issues to face, as the global energy system changes.

Global energy resource trading

One fundamental issue is that energy resources are not evenly distributed globally. That has increasingly shaped global development and global trade, as well as political battles and wars. Certainly, the location of fossil fuel resources, oil in particular, has been a major factor in shaping recent world history. The same may soon be true of renewable resources but perhaps to a lesser extent.

As IRENA's Global Commission pointed out:

renewable energy resources are available in one form or another in most countries, unlike fossil fuels which are concentrated in specific geographic locations. This reduces the importance of current energy choke points, such as the narrow channels on widely used sea routes that are critical to the global supply of oil. Second, most renewables take the form of flows, whilst fossil fuels are stocks. Energy stocks can be stored, which is useful; but they can be used only once. In contrast, energy flows do not exhaust themselves and are harder to disrupt. (IRENA 2019c)

So we will no longer be trading in fuels like coal, oil and gas, avoiding the problems that has presented. Most of the renewable resources will probably be used directly, locally/ nationally, at least initially. Countries with good renewable resources will do well and, for example, in general there is more wind in the north and south, more solar in the sunbelt, but specific locational and local geographical features are all important, with patterns varying around the world, as we will see in chapter 7.

However, this energy will not all just be used locally. There will also be trading between countries, for example in green gases and electricity, and this could expand. Some of the big winners may be countries that can trade in green fuels like hydrogen and synthetic methane. More controversial is the trade in biomass. As I noted in chapter 3, the United Kingdom (and the EU) is importing substantial amounts of forest-derived biomass fuel pellets from North America for power generation, and there is a global trade in palm-oil-derived biofuels for vehicles. There is a risk that the lure of these lucrative markets will lead to an environmentally destructive focus on monocultural cash cropping for export by some developing countries. That could also apply in some contexts to green gas production based on biomass.

There are also energy transport implications. Shifting coal, oil and gas around the world in bulk has led to some significant environmental problems (including major oil spills) and it also uses energy. There may still be risks with long-distance tanking of hydrogen, or ammonia or methane made from it, by ship. However, there are some technical fixes that may help with the transport energy issue (and with cargo shipping generally), for example, most obviously using biofuels and synfuels instead of marine diesel. In addition, wind power can be used to augment tanker vessel propulsion, either by the use of giant kites to deliver towing power (SkySails 2019) or by fitting large vertical cylindrical Flettner rotors to ships, making use of the Magnus effect (Aggidis 2017). We could be moving to a renewably powered renewable fuel delivery system.

Shifting electricity around is the other big long-distance trading option, possibly the largest. While there is some cross-border trade in electricity, wider power trading is as yet rare, but that may change significantly as the use of renewables builds up. For example, countries with large energy storage facilities (e.g. pumped hydro) will do well, especially if these plants can be accessed by other countries to balance their own supplies. Some countries may also have regular surpluses of green power available for trading, including long distance via supergrids, for direct use and for phased balancing. That type of trade would seem to offer substantial energy stability gains and could grow significantly, though, if expanded, a new pattern of energy trading would emerge with new winners and losers. (See Box 6.1 for some examples in the European and North African/Middle Eastern context.)

Box 6.1 Global green energy trading – a MENA assessment

The idea of a pan-EU supergrid network includes the option of wider links across the Mediterranean Sea to North Africa, and also possibly to the Middle East, as initially promoted by the Desertec group (Desertec 2011). That would allow imports of electricity from large solar projects there and also possibly from wind projects. In a proposal from the Desertec Industrial Initiative (DII), a development consortium mainly involving German power and finance companies, around 15–20% of the EU's electricity could be provided in this way, helping to top up supplies and also balancing out variations in local energy availability in the EU. However, it would not just be an export-based system. Most of the electricity would be used locally at or near source in Africa or the Middle East, although there might be some bilateral exports across and within the Middle East–North Africa (MENA) region. Indeed, the Desert Power 2050 study produced for the DII suggested that, by 2050, this would be vital since energy demand would rise in many of the newly emerging economies in the MENA area, possibly more so than in most parts of the EU. Local employment needs would also rise, and meeting that need could be a key issue (Calzadilla et al. 2014).

Certainly, trading electricity from their indigenous solar resources would open up new economic opportunities for the MENA countries. At present, some of these countries are reliant economically on oil and/or gas exports. However, the solar resource in these desert areas is vast, more than enough in principle to meet the combined needs of the entire MENA area and the EU, if fully developed and shared via supergrid networks. Even if only partly developed, the Desert Power 2050 study noted, the supergrid links would enhance energy security across the entire EU–MENA area (DII 2012).

Not all countries in the EU–MENA area would be able to contribute or trade equally. The Desert Power 2050 report identified three classes of country, based on their renewable resources and their level of energy demand: those with extensive renewables ('super-producers'), renewable-scarce countries ('importers') and countries with balanced renewables and demand ('balancers'). Under their proposed scheme, while no country would be allowed to import more than 30% of its electricity (most already import up to 10%, some more), there would be clear gains for super-producers, who could sell their surpluses, and for importers, who could buy in electricity at less cost than from their indigenous renewable sources. Morocco, Libya and Algeria are seen as potential super-producers (of solar), along with Norway (of hydro). Germany, Italy and, to a lesser degree, France and Turkey were seen as likely to be net importers, due to their high energy demand but lower renewable resource base. Egypt, Saudi Arabia, Syria, along with Spain, the United Kingdom and Denmark, were seen as the main balancers, with good renewable resources (mainly solar and wind) but also high energy demand (Elliott 2016).

In the event, the Desertec proposal was not followed up (Deign 2014) and it did focus mainly on solar CSP in the MENA area, whereas large-scale concentrated PV (CPV) now looks more likely to prosper there, due to its falling cost. In addition, the prospects for wind are also looking good, with 23 GW predicted to be in place in the MENA area by 2027 (NEU 2018).

As noted in chapter 4, there may be issues with supergrid power trading. For example, in the context of the idea of using solar in North Africa to provide power for Europe, there is the risk of 'land grabs' and exploitative relationships between rich EU countries and poor African communities (Elliott 2013a). In theory, it could be a mutually beneficial trade, with money from the North invested in projects in the South, and in grid links to them, leading to local energy, economic and employment gains. However, its net impact would depend on the terms of trade and ownership rights in relation to the resource and the systems.

That is not just an issue for North Africa and the EU or, for that matter, European green power imports from what could become a solar-rich Middle East. The proposals emerging for supergrid links across Asia also illustrate some of the likely regional geopolitical issues there, as for example with the proposal to link major CSP and wind projects in Mongolia and the Gobi Desert with energy demand centres in China, South Korea and even Japan. China should be able to meet most of its energy needs from its own vast renewable sources and perhaps also export surpluses at times, although there could be benefits from imports from outside for balancing. However, the situation in South Korea and Japan is very different. They have major space constraints, Japan especially, with high population densities and fewer areas available for solar, onshore wind or biomass projects. So they would likely be net consumers of green power, imported via supergrids, which would also help with balancing.

The proposals for extended supergrids may be technically and economically viable (Bogdanov and Breyer 2016), but the geopolitics could become quite difficult. For example, in the Golden Ring north-east Asia-wide concept, linking China, Japan and possibly India, it would make geographic sense for west-to-east supergrid lines to cross North Korea on the way to Japan (Movellan 2016).

Links have also been proposed westwards from China to Russia (linking up to the latter's hydro resource), with obvious political implications and, perhaps equally contentious, there have been proposals for links between Asia and Europe. In the latter case, an EC Joint Research Centre study looked at possible HVDC grid routes west to the EU, linking the major wind resources in north-west China and Mongolia, and also large resources on the way across. The 'high' route, via Russia and Ukraine, was seen as politically problematic, the middle route, across the Near East, avoided Russia but would involve some costly undersea/lake cables, while the lower route crosses some tough terrain, and also Iran and Afghanistan (Morgan 2018). Plenty of geopolitical issues there.

Similarly, the idea of linking the EU up to Russia's extensive renewable resources is contentious. As I indicate in chapter 7, Russia has yet to develop this large resource significantly, hydro apart, but in the longer term, if it did and supergrid links could be made, it could become a major exporter of green power, for example from wind projects in the vast windswept steppes of Siberia. Kazakhstan also has a huge wind resource and could become another major player (Elliott and Cook 2018).

Links from Europe to North America are also technically possible and open up another set of geopolitical issues. One route would be between the United Kingdom and Canada via Greenland (Purvins et al. 2018). The United States might find that hegemonically challenging, and so might the EU with, post Brexit, the United Kingdom being outside the EU. The North American link could of course be routed to Ireland. Indeed, that might be the most direct route. Interestingly, in that context, although the United Kingdom already has undersea links to the European mainland and plans more, including to Ireland, Ireland has recently been planning its own direct links south to mainland Europe.

That small local spur apart, most of the links I have looked at so far are about east–west trade, but what about long-distance north–south links? They might be very relevant in Africa, linking up and balancing the extensive renewable resources, large and small, across this vast continent (Elliott and Cook 2018). Most of the existing grid system is in the south, whereas much of the large solar resource is in the north. Given regional political conflicts and the poor state of the grid system in much of Africa, for the moment, with many rural areas still being off-grid and maybe 600 million people having no or very limited access to power, the emphasis is on off-grid power and local mini-grids. So realization of the big north–south supergrid idea may be some way off.

Some say that, in any case, big new grid links are not what is needed and worry that the main impetus for them is to send power from a few large centralized projects to major centres of energy demand, bypassing most local energy users. With very large hydro projects, like the 40 GW scheme being developed on the Congo river, that is certainly a risk, but it is also true that at some point it would make sense to link local mini-grids into national and then international networks – though it may take time.

There may be similar issues in relation to major supergrids between North and South America. For example, given the specific geographical access problems there, some say that may not ever be viable on a significant scale (see Box 6.2).

Box 6.2 Linking North and South America

A study by LUT in Finland and a Brazilian academic has found that both North and South America can reach 100% renewables (as Mark Jacobson's and earlier LUT studies have shown), and that local area/regional grids can help with balancing and trade. However, north–south links would be harder, in part because the distances are too great and access link paths constrained but also since the major centres of population are not handily located. Distances on their own are not a major problem with low-loss HVDC, and it is clear that supergrid links can help stabilize power systems, reducing the need for local energy storage/curtailment. Nevertheless, their economics depend on there being a locational (as well as temporal) match between supply and demand.

It is complicated by the approach that has evolved in the United States, where there are few interlinks between the main regional grids systems, which cover very large multi-state areas. The lack of long-distance integration is being challenged, but in many US renewable energy studies 'integration' still just means local regional grid upgrades. Perhaps unsurprisingly, then, local storage is often favoured, and it is getting cheaper too.

That pattern shows up in the LUT study: local storage often looks easier. It is further complicated by the fact that the United States uses a lot of fluid energy, gas and oil, and since these fuels are usually only available from well-defined production areas, it has long pipelines for that or tanks it about, including from deliveries to seaports from abroad. These fuels can also be stored there. It is all a bit different from electricity.

The LUT study looks at shipments by tanker of liquefied natural gas (LNG) from the south to North America, and that might fit in better with the existing energy trading pattern than long-distance power transmission, though not maybe in environmental terms, even if it is synthetic gas (SNG) made from (Brazilian) biomass. Or even, the study suggests, in energy and economic terms: at least in theory, supergrid power links are better. Against that, as noted above, it is not always easy to get access, given the geography, and *offshore* supergrid links are just not viable over these huge distances. So the paper's one-line summary says, a little baldly, 'Long-distance transmission lines cannot compete with energy storage technologies.'

However, it is also clear that, in most situations, although storage is the default position (and is getting cheaper), local, regional, national and international interconnections still ought to trump all else. But perhaps, in terms of balancing, more so for east–west power transfers than north–south transfers: the power available via east–west links follows the sun on a daily basis, although very long-distance, north–south links can exploit the global winter/summer differences (Aghahosseini et al. 2019).

Given that the prospects for major new supergrids in some parts of the developing world may be limited, at least in the short term, the emphasis instead is likely to be mainly on developing renewable resources locally. Indeed, as noted above, some would say that pressure to go for supergrids reflects the wrong developmental model, one in which large centralized hydro or nuclear plants dominate.

That may well be a fair description of what has often happened in the past but hopefully it can be avoided in future. As I have argued earlier (in chapter 2), there is no question that smaller-scale locally developed energy systems have a key role to play, and that is clearly the case in parts of the developing world. However, as the systems develop, there will also be a need for grid links to allow for power trading and to aid balancing. That does not mean we have to have giant centralized plants, or that smaller local plants are irrelevant, or that local grids and national, or even international, grids cannot coexist. We will need all types of grid, operating at their different levels, linking plants of different (but not gigantic) scale, possibly in some cases also including green gas grids and gas stores.

Global technology, carbon and materials trading

The idea of global power grids, and even more so green gas grids, may sound some way off but, as I have indicated, plans for new power grid links are being made around the world and interest in hydrogen as a new energy carrier is growing. Power grids have an edge at present, and the potential for more is vast, with much interest being shown in developing the necessary technology and systems (Liu 2016). That is even more the case for renewable energy-generation technology. It is a booming market worldwide. Global investment in renewable power and fuels in 2018 totalled $289 billion, $305 billion if large hydropower is included (REN21 2019).

In terms of hardware exports, the leaders include Germany, China and, to a lesser extent, the United States. President Barack Obama once said that 'the country that harnesses the power of clean, renewable energy will lead the twenty-first century' (Obama 2009), and US equipment export sales were clearly expected to be one of the pay-offs. According to the US International Trade Administration, by 2016 US exporters were capturing 3.2% of the global renewable energy technology market, and that was expected to rise to 5.6% by 2017 (ITA 2016).

However, China was clearly also trying to move into the global market. In the late 2010s, it certainly did very well with its hardware sales, of PV cells especially, leading those of all other countries in terms of value added from manufacturing investment in clean energy technology (CEMAC 2017). The United States and China have increasingly become engaged in something of a technology-led contest for markets in this area, but with China taking the lead, even in finding markets for its low-cost solar technology in the United States. That has led to complaints about 'dumping' and unfair competition. 'Protectionist' reactions became even more apparent in the United States from 2017 under Trump, who has been imposing further import controls on Chinese solar equipment. Clearly, international technology trade opens up some large, essentially political, issues. While 'free trade' is often espoused as the ideal by right-wing economists, in practice tariff barriers are often imposed to protect national markets. We may see more.

Mention should be made in the international and regional trade context of the new area of, in effect, 'virtual' trading, that is in carbon credits. International, regional and national markets for carbon credits have been created by setting overall carbon emissions caps and then creating tradable credits for the amount of carbon production avoided by clean energy or offset by carbon removal projects.

However, while there was initially enthusiasm for carbon trading from the financial sector, who saw carbon markets as a possibly lucrative new growth area, there are issues. Markets for carbon credits may not always be well linked with what is actually happening on the ground. Once you create an abstract intermediary credit, then, like money, it can take on a life of its own, opening up the possibility of speculation against future trades and future carbon values. In theory, it is all meant to relate back to actual carbon-saving projects, but the connection can become tenuous, even corrupted. It can certainly be hard to manage reliably. For example, accreditation of claimed emissions avoidance can be difficult, and it can be disproportionately costly and time-consuming to assess small projects in remote locations. So full accreditation assessment may not always be done, or at least projects may not be *independently* checked *on site* regularly, their carbon savings simply being deemed to have been delivered based on (paper) compliance reports. There have been anecdotal accounts of projects that were not running as claimed, or even projects that were accredited on paper which did not actually exist. That may be just hearsay but, given that very large sums of money can be involved, there is an obvious need for careful policing, which makes the whole process slower, more expensive and more bureaucratic.

I will be looking at some of the problems that emerged with emissions trading systems in overall global policy terms later in this chapter but, while some progress has been made with specific schemes and projects, it has been slow, as has the spread of carbon-trading systems. The United States toyed with the idea of a national 'cap and trade' system but decided not to go ahead, and, in general so far, carbon trading in its various guises has not been spectacularly successful in terms of supporting emissions reductions beyond those which might have happened anyway.

Nevertheless, interest in carbon pricing and trading systems has remained in the United States and elsewhere (Rosenzweig 2016). With heightened concern over climate change, we may yet see the adoption of 'carbon tax' polices in some form in the United States, although that would presumably have to be under a new president. However, cap and trade systems might be seen as a market-based alternative to the more interventionist Green New Deal proposed by some left-leaning Democrats and so might appeal more to the political right, despite their more usual stance of opposition to taxes. Interestingly, China has tentatively entered the carbon-trading arena, opening up an *internal* carbon market, which, given China's size, could easily rival the EU emissions Trading System (EU ETS) in scale, and China might, at some point, go global on this.

In the EU, despite the problems which I look at below, carbon-market trading is still seen as a key climate policy tool, and certainly the imposition of tighter national carbon caps could enhance the carbon market. So too could the development of negative emissions technologies, CCS and so on, and although, as I indicated in chapter 3 and in Box 4.5, that approach to carbon removal may have technical and economic limits, any carbon saved could presumably be included in trading systems.

Another area of more conventional trading, and one that could be increasingly crucial globally, concerns the need for rare materials for use in renewable energy systems. As noted earlier (Box 5.2), the expansion of renewables and battery storage systems is leading to growing demand for rare earth materials, as well as for lithium. These materials are mostly only available in a limited number of locations around the world. So it is likely that countries with these resources will seek to exploit their advantage to the full, seeking maximum possible income (Barron 2018). Indeed, there have been fears that limiting access to these resources could be used as a political weapon, with much concern expressed about China's alleged dominance in some areas.

This risk may have been overstated and is not new. Conventional energy systems also sometimes use these materials and 'rare earth' materials are not actually rare; they exist in many places, although often in very low concentrations. Demand is rising, but new finds and new sources and substitutes are possible, stimulated by the demand (IRENA 2019c; Overland 2019). Some argue that, as essentially commercial issues, it ought to be possible to resolve them. It is all part of global market rivalry. Certainly, you should not complain about that if you believe that market competition is the best way to drive technology forward.

Others say that the effective development of energy technology in this specific area, and in others, may require intervention in markets so as to avoid potentially corrupt practices and ensure fair competition, as part of a process of setting trade rules, and even as part of the wider context of global agreements on environmental protection and global climate change-related actions and policies. Given that the latter have arguably been of somewhat limited success so far, that may be optimistic. However, there have been some successes at the global level and, clearly, international agreements on energy and climate policy will be vital if we are to cut emissions effectively, equitably and fairly.

Global initiatives

Global responses to the threat of climate change have usually focused on the development of treaties and protocols at the international level, often based on global targets for greenhouse gas/carbon emissions reductions. The Kyoto Protocol, agreed in 1997, was a legally binding agreement under which industrialized countries were required to reduce their collective emissions of greenhouse gases by 5.2% from 1990 levels during the period 2008–2012. That may sound a small cut but, compared to the emissions levels that would have been expected by 2010 without the Protocol, this target represents a 29% cut. National targets were also agreed and ranged from 8% reductions for the European Union and some others to 7% for the United States, 6% for Japan and 0% for Russia, and permitted increases of 8% for Australia and 10% for Iceland.

Unfortunately, not all of the initial signatories ratified this agreement, notably the United States, the largest single source of emissions at that time. The United States had clearly been unhappy with the Kyoto Protocol, and, when George W. Bush was elected as president, its opposition became more forthright. In March 2001, Bush commented:

> I oppose the Kyoto Protocol because it exempts 80% of the world, including major population centers such as China and India, from compliance, and would cause serious harm to the US economy ... there is a clear consensus that the Kyoto Protocol is an unfair and ineffective means of addressing global climate change concerns ... I do not believe that the [US] government should impose on power plants mandatory emissions reductions for carbon dioxide, which is not a 'pollutant' under the [US] Clean Air Act. (Bush 2001)

Subsequently in 2002, the United States withdrew formally. Australia also refused to ratify the treaty although, by the time of the Earth Summit on Sustainable Development in Johannesburg in 2002, it was clear that most of the rest of the countries in the world would ratify it.

A similar pattern emerged more recently in relation to the so-called Paris Agreement, initially agreed in 2015, calling for signatories to work together to keep the global temperature rise well below 2°C above pre-industrial levels and to pursue efforts to limit the temperature increase even further to 1.5°C. The United States initially signed up to it, but under Trump it has sought to back out (UNFCCC 2016).

These and other protocols and agreements were developed under the UN Framework Convention on Climate Change in line with the scientific work of the Intergovernmental Panel on Climate Change but, although in theory legally binding, there were only relatively weak penalties for non-compliance.

Some however were hopeful that the introduction, as part of the Kyoto agreement, of schemes like the Clean Development Mechanism (CDM) would provide a positive incentive to investment in clean energy projects in the developing world. Companies in industrial countries that invested in such projects could claim emissions credits against the countries' national targets. It was a form of emissions trading, creating market value for emissions avoided.

Unfortunately, as noted earlier, in general the carbon cap and emissions trading approach has not worked too well. In some cases, perverse situations emerged. For example,

Russia's economic and industrial slowdown, following the collapse of the Soviet Union in 1991, meant that, despite its very heavy reliance on fossil fuel, its emissions were well below the 1990 baseline set in the Kyoto Protocol, and the zero-reduction emissions target it was given. So it had plenty of emissions credits.

That was a short-lived issue (Russia's economy recovered), but similar problems emerged subsequently with the EU Emissions Trading System (EU ETS). The value of carbon in the EU's carbon-trading market is in effect set by EU-determined national emissions caps. However, they were set at relatively high levels after resistance to proposed tighter caps from EU countries that were heavily reliant on coal use. So the tradable value of carbon was low and the carbon market has not proved as effective as hoped at stimulating new low- and zero-carbon projects.

Attempts to impose tighter emissions caps to increase the value of carbon credits and other policies aiming to speed up the development of low-carbon energy have continued to be fought off politically by countries still reliant on coal. A recent example of this has been in relation to the European Commission's proposal for a 2050 'net zero emissions' target, which needed a unanimous vote. It was vetoed by Poland, along with Hungary, Estonia and the Czech Republic. They blocked the commitment to decarbonization by 2050, preferring a vaguer time commitment (Morgan 2019). The final document removes the date and just says that the EU would 'ensure a transition to a climate-neutral EU in line with the Paris Agreement'. But some say that this fudge will undermine the Paris Agreement and also the carbon market (Keating 2019).

As can be seen, it has proved hard to overcome resistance to change by those with a vested interest in avoiding it. That includes not just those reliant on coal use and export but also some of the oil-producing countries. Coal and oil geopolitics, along increasingly with gas geopolitics, may thus inhibit the adoption of clean alternatives, at least for a while.

However, this situation should not be portrayed just in terms of simple resistance to change by groups with vested economic interests. That obviously exists, but there is a wider set of issues concerning socio-economic development and social equity. Some developing countries will argue, quite reasonably, that the old industrial countries have had the benefits of fossil fuel use for hundreds of years, which is one reason why they are rich. Why should newly developing countries be denied access to that route by punitive emissions caps?

The counterargument is that all countries will suffer from climate change and that the fossil route is not viable for the future for any of them. Moreover, although it is true that the industrial countries were once the largest emitters, now some of the newly developed countries have taken over, for example, China and to a lesser extent India, though the United States is still number two (GCP 2020).

One response to that from the poorer developing countries is to demand aid to help them to develop clean energy systems and, more immediately, to help them cope with climate change impacts, which could be increasingly costly for them. Aid was promised in the 2016 Paris Agreement but, predictably, the levels and payment contributions have been politically contentious. Trying to devise equitable solutions in a grossly unequal world is not easy.

Some battle lines ahead

While reaction and resistance to change may dominate at the global level, and need to be fought, there may also be things that can be done, and will be done, nationally and regionally, which can have a global impact. Climate change will be one driver but, as I argued in chapter 1, there are also economic incentives for change, quite apart from climate change or fossil fuel cost and scarcity.

That may even apply to oil-producing countries. A famous quote attributed to Sheikh Zaki Yamani, a Saudi Arabian oil minister three decades ago, was that 'the Stone Age did not end for lack of stone, and the Oil Age will end long before the world runs out of oil.' The reality is that it may take some time for the oil producers to make the transition but there are already some incentives at the margins. Growing demand for power for air conditioning and desalination of seawater in some Arab oil states is diverting some of their valuable oil from export. So they are investing in solar PV to meet this demand and also in CSP to provide power for oil processing, small steps in a better direction, which may yet be followed up more seriously. Indeed, at one stage, the Saudis were talking of a giant 200 GW solar programme, although that now seems to have been cut back to 58 GW by 2030, nonetheless a significant step (DW 2018). They do seem to be aware that relying on oil indefinitely is not an option in resource terms nor in terms of climate change: they have to diversify.

Significant steps are also being made in Africa, as I reported in my recent co-authored book (Elliott and Cook 2018). Some of it is smaller-scale and incremental, with solar projects beginning to spread, helping to meet local needs, but some of it is of more strategic importance. South Africa, the largest economy in Africa, has turned its back on nuclear and coal expansion and is planning a major expansion of renewables, mainly on the basis of their lower cost. Coal currently provides around 90% of South Africa's power, so this is a major policy change. In 2018, its long-proposed 9.6 MW nuclear expansion programme having stalled, the government decided instead to back a renewables expansion programme, with over 16 GW set for 2030, including 8.1 GW of wind, 5.7 GW of PV and 2.5 GW of hydro, so that renewables would then have a 36% share of overall power generation. The nuclear share would fall to 2%, gas to 16%, and coal-fired generation to a 46% share in capacity terms, though that might still translate to nearly 64% in power-use terms (Burkhardt and Vecchiatto 2018).

However, this new policy was hard fought, and it proved difficult to resist some revisions. While the commitment to renewables has remained in place, some limited coal expansion (1.5 GW) was later agreed, with the 2030 target raised to 59%. It was also pointed out that, under the initial agreement, it was always envisaged that some new nuclear plants would be considered later on, and in 2019 up to 2.5 GW was proposed, possibly small modular reactors but not until after 2030.

The potential for energy change in South Africa is very large (Bischof-Niemz and Creamer 2018), but the 2018 policy change was clearly not without its problems. Coal mining and the coal-fired power plants run by the state-owned power company are major employers in South Africa. There was some hostility from trade unions, who were worried about their jobs and none too happy with some of the renewables projects that had been established, based on private finance and independent power companies (Elliott and Cook 2018). A shift to renewables requires careful attention to issues like this or it may be opposed. In theory, replacement jobs and more should result from the switchover, but that may take time and there are issues of retraining to consider, not always an easy option for older workers.

More positively, much of the aid being provided to Africa for renewable energy development, for example by the EU, is now being recast with job creation in mind. The EU has a vested interest in this. It hopes that creating new local jobs, for young people in particular, might stem the flow of illegal migrants to Europe (Elliott and Cook 2019).

Employment concerns have also been an issue in the United States, where initial Green New Deal proposals met with opposition from some trade unions worried about job security (Marinucci and Kahn 2019; Roselund 2019). The Green New Deal proposal then produced by Senator Bernie Sanders, which looked to meeting 100% of US power and transport needs with renewables by 2030 and then moving on to full decarbonization by 2050, stressed that the plan was for displaced workers in the fossil fuel and other carbon-intensive industries 'to receive strong benefits, a living wage, training, and job placement'. Indeed, it seems to bend over backwards to provide support: 'We will guarantee five years of a worker's current salary, housing assistance, job training, health care, pension support, and priority job placement for any displaced worker, as well as early retirement support for those who choose it or can no longer work' (Sanders 2019).

Whatever form initiatives like this finally take, and there have been similar proposals elsewhere in the United Kingdom and globally (Saltmarsh 2019; Varoufakis and Adler 2019), it will be vital that they pay careful attention to the transitional employment issues and also to long-term employment implications. Labour's UK Green New Deal plan said it aimed to create guaranteed work in the new zero-carbon economy 'for those whose current roles are set to change', and it stressed the need for a 'just transition' and for 'a sustainable future based on good, secure, unionised jobs' (Labour GND 2019).

While it is true that there will be plenty of jobs in renewable energy, with more than 10 million people already employed in renewable energy globally, a key point is that what also matters is the *types* of job. It would be relatively easy to create many low-paid, low-skilled jobs, for example in biofuel production, but that may not be what is needed or wanted. Not all the jobs will necessarily be high-skilled, but it is argued that the 'just transition' must include the creation of good, worthwhile jobs and environmentally sound work (Elliott 2015).

A linked issue concerns the strategic focus of efforts to respond to climate change. In the short term, whatever else is done, there will be a need for emergency ameliorative measures to deal with impacts, along with longer-term adaptation to climate change where possible. Some of that will create employment and boost the GDP, and may also save money by reducing damage and social costs, at least in the short term (Farand 2019). However, it will not deal with the fundamental climate problem, which will get worse if emissions continue to rise. Indeed, some adaptive responses, for example using energy and materials to build more dykes, can lead to more emissions.

The term 'mitigation' is usually used to refer to measures aimed at avoiding or reducing carbon emissions, which will ultimately reduce impacts. However, finding the financial resources to do that can be a problem, especially for poorer countries. By contrast, adaptation can often deal with *local* impacts in the short term, and is likely to be cheaper than mitigation, for example through the development of renewables. If so, then, adaptation may be all that poorer countries, hard hit by climate crises, can afford, since mitigation is in any case unable to reduce local climate-related problems *quickly*. So mitigation may be deferred, leaving emissions reduction for others to deal with. This may be understandable but, given that there will be competition for funding for climate-related actions, if a focus on adaptation occurs on a wide scale globally, that could reduce overall emissions-reduction efforts, these being the only long-term *global* solution to climate problems (Schumacher 2019a).

More generally, this issue can be seen in terms of social equity conflicts. As Schumacher argues, 'investments in emission reduction benefit everyone while adaptation only benefits the party that undertakes it' (Schumacher 2019b). Put another way, local short-term concerns and global long-term concerns may differ.

That conflict links into the much broader issue of the overall direction taken by countries and by the world as a whole. Will there be a move to embrace environmental sustainability and a sense of global responsibility or a retreat into local economic protectionism and political isolationism? One recent energy geopolitics study set 'dirty nationalism' against a 'Big Green Deal', with 'muddling on' as a no-change alternative and 'technological breakthrough' as the wild card (Bazilian et al. 2019). But are those caricatures actually realistic or mutually exclusive?

A perhaps more conventional view is that, as a Russian analyst has put it:

> some countries will see the shift to the low-carbon future as a loss of their existing international competitive advantages. Others will see it, in contrast, as an opportunity to gain new competitive advantages by creating new national competences ... and to grasp/win new competitive niches in the new growing ... innovative markets being developed with the transition of the global economy to the low-carbon development path. (Konoplyanik 2019)

Russia may fall into the first category, as may some other countries that are heavily reliant on fossil fuel use and/or export. However, although it may take time, quite apart from suffering from climate impacts, if these countries do not change, then, as new options boom, they could find themselves trapped in decline with stranded assets and resources and technology no one else wants.

Political futures

There is a need for change, as the quite grim 2019 report from the World Economic Forum argues. It says that, while some progress has been made, few countries are ready for the transition, and it calls for 'swift action' (May 2019). As has already been argued, making that happen on a global scale will be hard, given the entrenched power of the fossil fuel industries, and some governments may find it hard to make the necessary changes.

However, on the positive side, plans for expanding renewables are popular with the public. A Green Energy Barometer survey for Orsted of 26,000 people in 13 countries found that 82% thought it was important to create a world fully powered by renewable energy (see Box 6.3).

Box 6.3 Public attitudes to renewable energy

Orsted survey respondents in support of a world fully powered by renewable energy:

China	Taiwan	Germany	Canada	Denmark	France
93%	89%	84%	84%	83%	83%
US	UK	Netherlands	Poland	Sweden	S. Korea
83%	82%	81%	80%	80%	77%
Japan	Average				
73%	82%				

Overall, 70% thought their country should be ambitious about green energy. China led at 89%. Of global respondents, 73% felt that green energy would boost economic growth, 85% thought their country should phase out coal, 47% wanted less use of nuclear, 26% wanted more, 19% felt it was about right now, 8% didn't know, 80% wanted more solar, 67% more wind offshore, 64% more onshore. More than 70% in China and the United Kingdom backed tidal power. Only 51% backed sustainable biomass but in China this was 74%. Only 37% backed natural gas (Orsted 2017).

A more recent study reported even stronger commitments to renewables in the EU: 92% of respondents said it was important for their national government to set ambitious targets to increase renewable energy use by 2030, up 3% on the results of a 2017 survey (EC 2019).

Some are nevertheless fearful that it may be hard to achieve the rapid transition necessary, while adhering to democratic processes. Others say democracy will help and needs to be expanded, not avoided, with full political engagement being vital (Willis 2019). Certainly, some of the actions needed can be voluntary and local. For example, in addition to the idea of citizen assemblies to thrash out views, it is sometimes argued that community-level activities can provide a helpful focus and that local energy projects can generate not just energy but also regenerate, mobilize and inspire communities (Brummer 2018; Koirala et al. 2016).

However, there can also be problems: getting the right balance between parochial and global concerns, 'thinking globally and acting locally', may not always be easy. Indeed, some say that there can be problems with too much decentralization. For example, a proposal by the Labour Party for the public ownership of the UK power system saw a key role being played by local energy co-ops and community groups but argued that:

> communities are not always open and democratic, and benefits and power can accrue to those who already have more wealth or time. For example, there is a risk of creating gated energy communities or 'local energy islands', where communities with the financial and physical resources to generate and supply electricity opt out of energy networks, leaving poorer communities with the disproportionate burden of financing wider infrastructure. Public ownership is thus required as a backstop to community control, to ensure that decentralisation reinforces rather than undermines shared regional and national infrastructure, and allows for the pooling of resources needed to guarantee universality of supply most efficiently. (Labour 2019a)

Not everyone will agree with that, or the idea that governments must intervene and reshape markets (Partington 2019b). Certainly, some pathways may not be politically palatable to all. For example, the global energy geopolitics study mentioned above noted that 'many Western policymakers assume that technological progress is best achieved in a liberal market underpinned by free trade. This is not necessarily the case. China has scaled up renewable energy through top-down rule and state planning' (Bazilian et al. 2019).

More prosaically, the messy world of actual politics and conflicting ideologies often requires pragmatic compromises and trade-offs to resolve differences, but that is not always possible: some trade-offs may not actually be acceptable. For example, Ofgem, the UK energy market regulator, has claimed that some green energy projects were getting unfair preferential treatment by not paying their full balancing costs, an argument I mentioned earlier in relation to so-called market profile costs (see Box 2.2). It was claimed that this special treatment undermined system stability and market economics and also passed higher cost ultimately on to consumers. In response, Exeter University Professor Catherine Mitchell and her team insisted that a false trade-off between costs, energy security, social equity and environmental sustainability was being made: 'we have moved on from trade-offs between sustainability and equity or equity and security' (Mitchell et al. 2019).

If we want to change the energy system and move towards sustainability, then new frameworks and new structures will be needed, and there will be new trade-offs and balances to make, along with new definitions of social and ecological equity (McCauley 2019). Some say this means moving beyond capitalism and its competitive growth-reliant ethos. No one as yet has necessarily come up with the right formula (Green 2019; Monbiot 2019) and some ideas seem fantastical (Dale 2019). However, one thing is clear: those with vested interests in the existing system are unlikely to be willing to help much.

Despite the political uncertainties, some are nevertheless optimistic about the future. For example, IRENA's Global Commission has produced a 'New World' report on the geopolitics of the energy transition which says that, 'Just as fossil fuels have shaped the geopolitical map over the last two centuries, the energy transformation will alter the global distribution of power, relations between states, the risk of conflict, and the social, economic and environmental drivers of geopolitical instability.' Moreover, it claims that renewables will be 'a powerful vehicle of democratization because they make it possible to decentralize the energy supply, empowering citizens, local communities, and cities' (IRENA 2019c).

We shall see! There is no question that the world will change as and when the energy transition spreads. Summarizing the likely trends, a recent study concluded that:

because renewable energy resources tend to be more evenly distributed geographically than are fossil and nuclear fuels, the economic and security advantages of access to energy will be more evenly spread among countries, there should be fewer risks related to transportation chokepoints and less reason for great powers to compete over valuable locations. In sum: international energy affairs will become less about locations and resources, and thus less geopolitical in nature. (Overland 2019)

However, that does not mean there will be no geopolitical issues. The study added that, 'as renewable energy resources are abundant but diffuse, technologies for capturing, storing and transporting them will instead become more important. International energy competition may therefore shift from control over physical resources and their locations and transportation routes to technology and intellectual property rights.'

With that prognosis in mind, in the next chapter I will look at what has been done and planned so far around the world. It is a mixed story, with some notable successes but also some laggards. One conclusion might be that it will not be possible to get all the way to 100% renewables as quickly as some hope, certainly in terms of energy as opposed to just electricity. Indeed, some would say that, realistically and pragmatically, there is no need to get to *exactly* 100% – and that there may be some end uses in some countries that cannot be easily or quickly decarbonized. However, a more positive view is that, although it may take time, progress can be made on all fronts, given that the main constraints are political, not technical, and policies can change.

7
Global Action

Renewables are expanding rapidly on a global basis, but the pattern is uneven. Some countries and regions are doing better than others. This chapter looks at what is happening around the world and at some projections for the future.

Options for global energy change

The planet is in trouble, with carbon and other emissions threatening to destabilize the global climate system. This chapter looks at examples of what is actually happening around the world in response, focusing on the various national renewable energy programmes, including in China, which is in many ways the current renewables leader, certainly in terms of installed capacity. IRENA's Global Commission on the Geopolitics of Energy Transformation says that 'no country has put itself in a better position to become the world's renewable energy superpower than China' (IRENA 2019c). However, it notes that many other countries that are heavily reliant on fossil fuel use and imports also have much to gain in economic and strategic terms from developing renewables.

That said, the situation, in terms of future prospects, is far from ideal. Some countries may find making changes a challenge (heavy coal users and oil producers especially), and globally energy demand continues to rise. In its 2019 review, the Renewable Energy Network for the 21st Century (REN21) says that, in 2018, global energy demand increased by an estimated 2.3%, the greatest rise in a decade. This was due to 'strong global economic growth (3.7%) and to higher heating and cooling demand in some regions. China, the United States and India together accounted for almost 70% of the total increase in demand. Due to the rise in fossil fuel consumption, global energy-related carbon dioxide emissions grew an estimated 1.7% during the year' (REN21 2019). While renewable energy use has continued to expand (see Box 7.1), REN21 says it is not expanding fast enough to cut these emissions.

Box 7.1 Global renewables growth

As of 2017, REN21 says renewable energy accounted for an estimated 18.1% of total final energy consumption globally (TFEC). Modern renewables supplied 10.6% of TFEC, with an estimated 4.4% growth in demand compared to 2016. Traditional use of biomass for cooking and heating in developing countries accounted for the remaining share. The greatest portion of the modern renewables share was renewable thermal energy (an estimated 4.2% of TFEC), followed by hydropower (3.6%), other renewable power sources including wind power and solar PV (2%), and transport biofuels (about 1%).

In the power sector, REN21 says renewables are increasingly preferred for new electricity generation. Around 181 GW of renewable power capacity was added in 2018, setting a new record just above that of the previous year. Overall, renewable energy now accounts for around one-third of total installed power generation capacity worldwide. Nearly two-thirds (64%) of net installations in 2018 were from renewable sources of energy, marking the fourth consecutive year that net additions of renewable power were above 50% (REN21 2019).

As I noted in chapter 6, there is clearly a problem. REN21 says that: 'While there has been much progress on renewables, energy efficiency, and access to electricity and clean cooking facilities over the past decade, the world is not on track to meet international goals, most notably limiting the average rise in global temperatures to 1.5°C as stipulated under the Paris Agreement.'

It warns that, in terms of energy:

> the overall share of renewable energy (both modern renewables and traditional biomass) in TFEC has increased only gradually, averaging 0.8% annually between 2006 and 2016. This modest rise is due to a negligible change in the traditional use of biomass coupled with overall growth in global energy demand since 2006 (annual average increase of 1.5%). Despite strong demand growth in modern renewables, especially renewable electricity, these two factors have slowed gains in the combined share of renewable energy in TFEC.

REN21 is not alone in warning that progress on energy, as opposed to just electricity, is too slow. As I have noted, BNEF, IEA and others have said the same. Energy demand is booming so much, especially in transport, that emissions savings in others sectors, and through the spread of renewables, are being overwhelmed. What can be done?

In terms of what has actually been done, so far attention has often been focused on the easiest options, mostly in the power supply sector, but now a sense of urgency seems to be emerging: the focus for the energy transition must widen. For example, the CEO of the UK's Climate Change Committee has said, 'If there was ever an idea that we could approach this as a "sequential" transition – moving from Power, to Transport, to Heat, to Industry and Agriculture – then that thought needs to be re-examined. Tougher targets imply a different kind of sectoral strategy. Bluntly, we will need to move quickly to decarbonise every sector in unison' (Stark 2019a).

Ramping up renewables faster in all sectors will help, but we also have to get demand tamed and emissions cut back. Some, logically enough, want to focus on transport, but that is maybe the toughest nut to crack. Some progress is being made, with plans for banning fossil-fuelled cars in some countries, but, if the result is just more EVs, that may not be sufficient. Private cars are much less energy efficient than public transport, and do we really want to use precious green power to keep cars running? We need changed transport policies.

Given that this may take time to have an impact, focusing on reducing domestic heat loss is a potentially easier and faster option for carbon saving in many countries. Green heat supply might also make sense, using solar energy especially, biomass maybe less so, given its eco/land-use issues, although it is more efficient when used in CHP plants linked to heat stores and district heating networks. However, as I have noted, many plans at present look to using green *power* to run heat pumps: but will there be enough for that and for EVs?

There is thus some uncertainty as to which way to go in terms of optimal and rapid emissions savings. As I noted in earlier chapters, the beleaguered nuclear industry may see the uncertainty as an opportunity to get back in the game, offering power but also maybe heat and hydrogen. In parallel, the fossil fuel lobby looks to CCS, or CCU, and maybe wider geo-engineering options, to let them stay in business.

However, as I have argued in this book, the alternative pathways, based on reducing demand through improved efficiency and on renewables as the supply-side driver, seem to offer a much better future, although the necessary changes have to be made rapidly and extensively. This chapter looks at what is being done around the world and at the prospects for the future.

Europe

Europe initially took a lead in the development of renewables, with pioneering wind projects in Denmark and then Germany, backed increasingly by positive European Union (EU) policies on climate change. The EU (including the United Kingdom) played a major role in shaping the Kyoto Protocol and then the 2016 Paris Agreement.

The EU currently looks likely to just about meet its overall target of getting 20% of its energy from renewables by the end of 2020. While some member states may fall short (notably the United Kingdom, Ireland and France), others have exceeded the renewable energy targets they were given, with for example Sweden already at around 54%, although it has been beaten by Norway and Iceland (both around 70%), both of which are outside the EU. For the next phase, a 32% EU-wide renewable energy target has been set for 2030, along with a 32.5% by 2030 energy-efficiency improvement target. Longer term, the EU wants to have 'net zero' emissions by 2050, although, as noted earlier, not all member states have signed up to this date.

Most of the lead countries, in terms of renewables, have large hydro and/or biomass energy resources, but even those with less of these advantages have done well, most notably Germany, the largest economy in the EU. It has expanded renewables dramatically, and by 2018 had 120 GW in place, including 49 GW of wind and 46 GW of PV solar. Its current target is to get 65% of its electricity from renewables by 2030, on the way to at least 80% by 2050. In terms of total *energy*, it is looking to a 60% contribution from renewables by 2050, with energy demand cut by 50% from what it would otherwise have been.

Germany's strong commitment to renewables was linked to the decision, reinforced by the Fukushima accident, to phase out nuclear power. The last of its nuclear plants is scheduled to close in 2022. That raised concerns about likely disruption, although so far renewables expansion seems to have been able to more than compensate (see Box 7.2).

Box 7.2 Nuclear phase-out and emissions rises in Germany

Critics initially argued that phasing out nuclear would inevitably lead to increased use of fossil fuels and to more emissions. However, according to a World Nuclear Industry Status Report (WNISR), that has not been the result. It noted that, between 2010, the year prior to the shutdown of the eight oldest nuclear plants, and 2016, 'the increase of renewable electricity generation (+84.4 TWh) and the noticeable reduction in domestic consumption (−20.6 TWh) were more than sufficient to compensate the planned reduction of nuclear generation (−56 TWh), enabling also a slight reduction in power generation from fossil fuels (−13 TWh) and a threefold increase in net exports' (WNISR 2017).

But what about emissions? That is more complicated, due partly to the power exports that WNISR mentions. Although some fossil plant output is being used to balance renewables, the rapid expansion of marginal cost renewables has pushed fossil plants out of the domestic market to some extent, with their surplus power being sold, very lucratively, to France and other neighbours. The Greens in Germany want to cut back on coal use, and that has been agreed long term (Kirschbaum 2019). However, meanwhile, continued coal use means that emissions from the power sector have not fallen as much as hoped, only by around 4.1% in 2018, and that was offset by a rise in emissions from the transport and industry sectors, so that total emissions only fell by 0.5%, despite total energy consumption falling by about 5% (Platts 2018; Wettengel 2018).

Nuclear phase-out commitments have also been made in Belgium, Spain and Switzerland, while after Fukushima Italy voted overwhelmingly in a referendum not to go nuclear, a position already adopted by Denmark, Austria, Ireland, Greece and Portugal. France plans to cut back its heavy reliance on nuclear to 50% (of power) while expanding renewables, aiming to at least treble its renewable capacity by 2030. Some countries in the eastern EU, with old Soviet-era plants, remain committed to nuclear, but most are also expanding renewables significantly. The same is true for the United Kingdom, making it one of the few European countries still seeking to significantly expand both nuclear and renewables. It should be noted that the devolved government of Scotland does not support new nuclear and has pushed renewables very strongly, meeting more than 60% of its current power needs from them compared to around 33% in the United Kingdom as a whole.

In terms of overall renewable developments in the EU, as mentioned in chapter 2, progress has been somewhat slowed by the shift in Germany and elsewhere from guaranteed price feed-in tariff (FiT) support systems to competitive market-capacity auctions. That may have reduced costs to consumers but it has also reduced the amount of capacity going forward. The situation varies throughout the EU, but in Germany, coupled with the continued use of coal (until 2038), the funding cutbacks could mean, as a paper from Yale University argues, that German carbon emissions will fall by only 62% by 2050, well below its 80% emissions-cut target, transport being a key problem (Hockenos 2018).

However, as I noted in chapter 2, the influential German BDI industrial group has outlined plans for how Germany might get back on track, even looking at a 95% emissions-cut scenario (BDI 2018).

The BDI says that 80% is 'technically and economically feasible' but 95% 'would push the boundaries of foreseeable technical feasibility and current social acceptance'. While 'assuming optimal political implementation, the climate paths' macroeconomic effects would be neutral to slightly positive, for an 80% ambition even without global consensus', it warns that the 95% path would need a global consensus since Germany could not go it alone.

Fortunately, in this regard, there are some positive signs from other EU members. Spain now reportedly plans to switch to 100% renewable power by 2050 and to completely decarbonize its economy soon after (Neslen 2018). Denmark is still making good progress. Consumption of coal fell by 25% in 2017, while consumption of renewable energy increased by just over 11% to around 55% (of electricity). Overall, energy-related CO_2 emissions continued to fall, by 38% since 1990, and it is aiming to be carbon free by 2050 (State of Green 2018).

These plans have been aided by the falling cost of renewables. In the Netherlands, new subsidy-free wind projects are emerging (Richard 2018b). Portugal has seen PV solar projects falling to a record low of €14.76/MWh and, as costs continue to fall, there are plans for subsidy-free PV projects even in the often cloudy United Kingdom, with the UK Solar Trade Association predicting that PV may get down to £40/MWh by 2030 (Stoker 2018). Onshore wind could reach that too, if the current, arguably rather perverse, UK government blocks on it were removed (Elliott 2019b). So, with offshore wind already booming and projects going ahead from 2025 at £40/MWh or less (New Power 2019), the United Kingdom could have well over 60 GW of renewables in place by 2030, maybe 100 GW or more, up from around 43 GW now. That would put it well on the way to its target of being 'net zero carbon' by 2050. I will be looking at some scenarios like that in the next chapter.

Grid development and grid balancing will be major issues for the future, as they already are in Germany, where most of the wind generation is in the north and most of the load in the south. The development of a north–south supergrid, and local grid reinforcement, has met with some local opposition. However, Germany is also developing a range of large-scale storage systems for grid balancing, including hydro pumped storage, power-to-grid hydrogen production, heat stores and battery systems. In parallel, as in the United Kingdom and elsewhere, there is much enthusiasm for local smart grid systems, Germany being especially well placed for peer-to-peer trading, given that nearly half of its renewable capacity is small-scale and locally owned by prosumers, local energy co-ops and community groups.

The United States of America

The United States may have been resistant to signing up to global climate change policies (see chapter 6) but, despite major cutbacks in support under President Trump, renewables are doing quite well, with over 245 GW in place, generating about 18% of its electricity, and expansion is continuing as costs fall.

Wind (at over 94 GW so far) and PV solar (at over 50 GW) are now seen as clearly economically competitive with coal (Mahajan 2018). Onshore wind is certainly doing well, supplying more than 6.5% of US electricity overall and more than 10% of total electricity generation in 14 states, more than 30% in Iowa, Kansas, Oklahoma and South Dakota in 2018. It is also proving to be popular, with a survey finding strong local support (Firestone and Kirk 2019).

So onshore wind seems likely to continue to boom, with 35 GW more in the pipeline (Patel 2019). A report for the US Department of Energy said that the national average price of wind power purchase agreements had fallen to around 2¢/kWh, and it looked to capacity additions of 8,000–11,000 MW/yr from 2018 to 2020, but with market contraction likely from 2021, as tax incentives were phased out (DOE 2018a). Despite Trump's cuts, solar is also doing very well, with over 188 are phased out (DOE 2018a). Despite.5 GW of utility-scale PV projects in the pipeline at the end of 2017 and, in all, more than 1.3 million distributed systems in place

(Berkeley Lab 2018, 2019) – a massive and continuing surge. It may be further boosted by California's new 100% clean power goal: it is aiming to get 60% of retail electricity from renewables by 2030 and 100% by 2045.

Next up, after a long delay and some local political opposition, offshore wind is finally taking off in the United States. According to the US Department of Energy, the country now has a project pipeline of more than 25 GW of capacity, with nearly 2 GW expected to be in place by 2023 to follow up the 30 MW already installed. Almost all of the expansion is set to be on the Eastern Seaboard, led by Massachusetts (DOE 2018b).

Longer term, a seminal report from the US National Renewable Energy Laboratory (NREL) found that the US renewable resource base was sufficient to support 80% renewable electricity generation by 2050, even in a higher demand-growth scenario. It also looked at a 90% option, with 700 GW of wind and solar PV (NREL 2012). More recently, a pan-US study from Stanford University, a subset of the already mentioned global '100% renewables' study by the same team, outlined a scenario for 100% renewables in 50 US states (Jacobson et al. 2015), with full grid balancing, including through the use of CSP heat stores and upgraded grids. Grid integration is poor in the United States, and several studies have pointed to the need for improvements in order to enable and manage 'high renewables' contributions 'by moving away from a regionally divided electricity sector to a national system enabled by high-voltage direct-current transmission' (MacDonald et al. 2016).

At present, there are no national renewable energy targets for the United States, although there have been independent proposals for 'high renewables' shares, and many action plans at state and city level. A study by leading consultants Wood McKenzie has suggested that getting to 50% (of power) nationwide would be relatively easy. Wood Mackenzie's director of Americas Power Research, Wade Schauer, says: 'Our analysis of the data suggests that reaching 50% of supply from intermittent renewables system-wide is relatively straightforward in most of the US.' But, he added, 'above 50%, integration challenges accelerate rapidly. Achieving full decarbonization will require long-duration energy storage, and the electric grid will need to roughly double its capability' (ESJ 2019).

That may still be a way off. In a scenario from the US Energy Information Administration, renewables are shown only supplying 31% of power by 2050 (EIA 2019). However, EIA estimates have been seen as rather conservative (Stark 2019b). Certainly, as indicated above, renewables are moving ahead quite rapidly, in part since they are becoming commercially viable. That, along with the advent of cheap shale gas and falling demand for electricity in some regions, has presented a major challenge to nuclear, with many older nuclear plants having to close and new nuclear projects failing. At present, only one new plant is under construction and, although there are hopes for new technology (e.g. small modular reactors), that is some way off. So, interest in new types of small reactor aside, nuclear seems to be on the way out in the United States (Millward 2018).

So what are the prospects for the future? With Trump in command, it is a little hard to predict. He is not fond of renewables, likes nuclear and coal, but in the real world, along for the moment with gas, it is renewables that seem to be winning in the United States as elsewhere. The United States' current policies on climate change may seem perverse and hard to fathom at times (Newman 2018) and its international stance more so, but its emissions have been falling, as it made clear at the 2019 G20 Summit in Japan (see Box 7.3).

Box 7.3 US policy position as reported to the Osaka G20 Summit

'The United States reiterates its decision to withdraw from the Paris Agreement because it disadvantages American workers and taxpayers. The US reaffirms its strong commitment to promoting economic growth, energy security and access, and environmental protection. The US's balanced approach to energy and environment allows for the delivery of affordable, reliable, and secure energy to all its citizens while utilizing all energy sources and technologies, including clean and advanced fossil fuels and technologies, renewables, and civil nuclear power, while also reducing emissions and promoting economic growth. The United States is a world leader in reducing emissions. US energy-related CO_2 emissions fell by

14% between 2005 and 2017 even as its economy grew by 19.4% largely due to the development and deployment of innovative energy technologies. The United States remains committed to the development and deployment of advanced technologies to continue to reduce emissions and provide for a cleaner environment' (*Japan Times* 2019).

Whether that carbon reduction will continue is unclear. Trump's national policies have had severe impacts on some renewable energy sectors, PV solar especially (Ellsmoor 2019b), but the wider picture is more complex. The United States is undergoing an energy transformation, shale gas and renewables replacing coal, so that its carbon emissions have fallen. Although that is obviously welcome (Shaw 2018), some would say that this was *despite* Trump. For example, it has been argued that most of the reductions were, in fact, due to the working through of President Obama's policies and that, with these now abandoned, emissions will rise (Jacobson 2018). Indeed, they seem to have done so, driven by rising transport demand (Irfan 2019).

Longer term, as noted in chapter 4, it may not be possible (or wise) to sustain the shale gas revolution: there are diminishing returns on well productivity and investment, and the environmental impacts of fracking can be serious, with some predicting a slowdown (Dyson 2019). By contrast, renewables look like a much better deal, and support for their expansion figures strongly in the various Green New Deal proposals (Amadeo 2019; Sanders 2019). That could involve significant federal and state government intervention.

As I noted in chapter 6, a more indirect and arguably more incremental market-orientated carbon-tax, carbon-pricing approach also has some backing (Teirstein 2019), though a survey has indicated that this was less popular than a direct government support-led regulatory approach (Fowlie 2019). The experience with emissions trading in the EU and elsewhere (see chapter 6) arguably suggests that the US survey respondents might be right, though with 'hypothecated' taxation and targeted re-spend on renewables, it might be better received (Burke and Byrnes 2019). The policy battle continues.

China

In terms of renewable deployment, China is leading the pack, with more renewable energy capacity installed (around 696 GW in 2018) than any other country (Probert 2018). There are some major projects, more investment having been made in renewables so far than by anyone else (Frangoul 2018). For example, in 2017, it invested $132.6 billion in clean energy, up 24% on 2016, setting a new record, and dwarfing the United States' investment of $56.9 billion (BNEF 2018b). It also planned to shut some coal plants and halt construction of new ones, in part as a response to the massive urban air-quality problems that emerged after its breakneck economic expansion, which was based mainly on coal. In 2017, the Chinese National Energy Administration claimed that 'between 2016 and 2020, we plan to halt construction or suspend building of new power plants with a total capacity of 150 GW, and shut down 20 GW of outdated capacity' (Reuters 2017).

In the event, it may not have lived up to that plan, there being reports of significant new coal-plant construction, including for projects that were supposedly abandoned. With demand rising, China clearly has problems, but it obviously wants to cut emissions and, to achieve that, in the short term some of its coal plants have been throttling back their operation part of the time (IRENA 2019c; Shearer 2019). However, longer term, it will have to move away from coal and certainly it is pushing renewables very hard.

Hydro remains the largest renewable (at around 350 GW) but, in terms of new renewables, wind has led the way, more than 184 GW having been installed by 2018 (GWEC 2018), although PV has been catching up, with more than 175 GW in place in 2018 (Baraniuk 2017). There are impressive solar projects on land and also on lakes (McPhee 2017). Its largest floating PV array so far is 40 MW, but a 150 MW project is underway.

However, renewable expansion has not been without its problems, which have led to the rate of growth for wind and also PV slowing from the earlier high levels (Ng 2017) and a fall in investment. One reason was that, as capacity was built up, there were continuing 'curtailment' problems. With wind, the main resource areas are often remote from load centres and grid links can be poor. With PV, although sunlight is ubiquitous, some regions get more than others. So in some cases local curtailment of output from both technologies can be very high. For example, the PV curtailment rate across China rose 50% in 2015 and 2016, as it failed to use more than 30% of available power in the north-west provinces of Gansu and Xinjiang. Curtailment of surplus wind output had reached 20% in 2016 nationally, much more in some remote locations with poor grid links and 43% in worst-case Gansu province.

As a result, the National Energy Administration (NEA) decided that, while grid and integration improvements were made, no new PV capacity would be added before 2020 in Gansu, Xinjiang and Ningxia provinces, and no new wind plants approved in Jilin, Heilongjiang, Gansu, Ningxia, Inner Mongolia and Xinjiang between 2017 and 2020.

With rapid expansion slowed and more attention being paid to grids, progress has been made on reducing curtailment. Reportedly, it fell by 15% nationally for wind in 2017, and efforts are being made to get curtailment down to 30% in the worst locations (Gansu and Xinjiang) and to 20% in Jilin, Heilongjiang and Inner Mongolia. The expectation was that it could be completely eliminated in Heilongjiang, Jilin and Ningxia, while Inner Mongolia was expected to reduce it to below 5% (Y. Liu 2018). According to IRENA, progress has continued with curtailment levels for wind farms dropping to 7% in 2018 from 13% in the previous year, while for solar PV plants they dropped to 3% from 5.8% over the same period (IRENA 2019d).

However, curtailment has not been the only problem to have emerged from the initial rapid expansion of wind and PV solar. It also led to major cost/subsidy overshoot problems. Indeed, the government was forced to suspend all new subsidized solar capacity approvals for a while, after a record 53 GW capacity increase in 2017 left it with a backlog of at least 120 billion yuan ($18 billion) in subsidy payments. So there was a big and sudden slowdown (Merchant 2018).

This move, and the curtailment-related slowdowns for PV and wind, have evidently helped with the subsidy-overspend issue, and that is also being helped by the fall in the cost of PV and wind. In January 2019, China's National Reform and Development Commission (NRDC) reportedly said that solar construction costs in China had fallen 45% from 2012 to 2017, while wind project costs had dropped 20% over that period. So the subsidy system was being revamped, with some wind and PV projects able to go ahead subsidy free. Some projects will still get subsidies, but the NRDC claims, 'The economic efficiency of projects has steadily increased, creating favourable conditions for state subsidies to retreat and pressures on subsidy funds to ease' (Reuters 2019a).

With the curtailment problem on the way to being sorted and the cash crisis contained, it seems likely that, despite the cuts and the fallback in investment growth, China's renewable power capacity will continue to rise. By the end of 2018, it had reached 728 GW, up 12% on the year before, according to the NEA, this representing 38.3% of China's total installed power capacity. It included an extra 20 GW of wind and 44 GW of PV. As Reuters says, 'China has tried to change the "rhythm" of renewable power construction to give grid operators time to raise transmission capacity and ensure clean electricity generation is not wasted.' The NEA noted that overall rates of waste in the wind-power sector had fallen to 7% in 2018, down 5 percentage points on the year before, although the major wind-generation regions of Xinjiang and Gansu in the far north-west had still failed to get around a fifth of potential wind power onto the grid over the period. So there is still a way to go (Reuters 2019b).

Further improvements are clearly needed. However, looking at it optimistically, a report from the US Brookings Institution believes that 'if renewable energy curtailments were to be resolved, then China's share in meeting new electricity needs will raise from 37.8% to 63.4%. Similarly, it will increase its share of total electricity generation by 1.6% (from 26.4% to 28.0%)' (Dong and Ye 2018).

In some ways, it is a little surprising that China has had these problems. With ostensibly high levels of central control, China ought, you might think, to have more coherent energy-system planning and regional grid coordination. The reality may be different. It seems that what exists is what has been described as 'fragmented authoritarianism'. There are, it seems, rival bureaucratic cliques with conflicting or at least confused jurisdictions. For example, edicts may come from central government about building more generation capacity, but the local regional agencies are responsible for local grids, and they may not always be given the resources to provide what is needed. Clearly, much more needs to be done in terms of national and local grid strengthening, better project siting and systems design, as well as more integrated planning and the development of better institutional policy processes (Huenteler et al. 2018; Ye et al. 2018).

Meanwhile, in terms of grid upgrades, one central government priority has been to improve links to the giant (and controversial) 22.4 GW Three Gorges hydro project, which is in the middle of the eastern part of the country, some way from urban centres of power demand which are on the eastern and southern coasts. A series of HVDC links have been built to East and South China over distances of around 1,000 km to transfer electricity from that plant. In all, the total capacity of the HVDC links is 7,200 MW, with line losses put at about 3%. That should also help wind and solar projects in the area. So will the much wider HVDC 8 GW capacity network that is being built across the country, with 30,000 km having been installed so far and plans for 90,000 km in all. In addition, there are longer-term plans for links to Kazakhstan, Mongolia and Russia (Fairley 2019).

While work on these links, and on renewables, continues, China is also still seeking to expand its nuclear programme from its 3.9% power contribution (from 35.7 GW of capacity) in 2017 up to around 6% with, at one point, a target set of installing 58GW by around 2020. However, there were delays (Green 2017; Kidd 2017) and although the programme is certainly very ambitious in global terms and is still progressing, it has to be put in perspective. Renewables, including hydro, already supply around seven times more power in China than nuclear, and its wind power output has overtaken that of nuclear, with wind power still expanding quite rapidly, along with PV.

The expansion of conventional nuclear may have slowed but there is progress with new nuclear technology and possible new applications. China's 250 MW HTR-PM high-temperature pebble-bed helium-cooled reactor (still being built) can, it is said, be used in combined heat and power mode to vary the ratio of electricity-to-heat output. Possible heat applications are seen as 'desalination of seawater for human consumption, production of hydrogen, or a wide range of other high temperature heat applications in industry' (WNN 2017).

So what is the bottom line for China? Not everyone likes China's politics and human rights record, but its green energy programme is impressive. Its overall plan is to get 20% of its energy from low-carbon sources by 2030, and renewables look likely to make up the bulk of that. They already supply around 25% of its electricity and about 12% of its energy, and the potential is vast. One study puts it at up to nearly 5 TW for onshore wind, 1 TW for offshore wind and 13 TW for solar PV. It also suggests that, as more capacity of each of these is added on a widely distributed basis, balancing will become easier (Hu et al. 2019).

However, that is far off. For the moment, while China does seem to be addressing its problems with curtailment and weak grids, it still has major issues with coal use. It is also true that China did not sign up to very stringent targets for carbon in the Paris COP 21 accord, which was one reason why Trump wanted out of the agreement since the United States had signed up to what he saw as unfair targets. The Paris Agreement target for the United States was a 26–28% domestic reduction in greenhouse gases by 2025 compared to 2005, making its best effort to reach the 28% target. China did however commit to a peak in CO_2 emissions by 2030, with best efforts to peak earlier. It also pledged to cut emissions per unit of GDP by 60–65% of 2005 levels by 2030, potentially putting it on course to peak by 2027. It could no doubt try harder, on coal especially: it is the largest coal user in the world and, despite efforts to cut emissions and demand, it is still expanding its use of coal, albeit at a slower rate, as demand rises.

This is not a problem unique to China (except in scale) and hopefully it will change, given the cost advantages of renewables and as the potential for demand reduction is realized. However, China's emissions performance so far would make it hard for it to play much of a global leadership role in climate policy. Indeed, despite the United States having rowed back from the international climate agreement, there were reports that China did not feel able to become a global climate champion (Haas 2017). Nevertheless, these days it is a major player in renewables, not just nationally but also in terms of its expanding trade and investment role overseas, for example in Africa. There, about a third of the new energy capacity has been led by Chinese funding, around half of it being for renewable energy projects. But China is not alone in this. Western countries and companies are also keen to invest in what could be a vast new market in Africa (Elliott and Cook 2018).

The rest of the world

China's recent focus on Africa highlights the possibility that some newly developing countries may soon become major players in the global economy and also in renewables. IRENA has said that Africa has the potential and the ability to utilize its renewable resources to fuel the majority of its future growth with renewable energy but, hydro and recent progress in South Africa apart (see chapter 6), it has a way to go. It is the same for South America, although some countries there are much further advanced, notably Brazil.

At one time, there was much expected economically of the so-called 'BRIC' group (Brazil, Russia, India and China). China has certainly delivered both in terms of economic growth and renewables; so has India, though to a lesser extent on renewables. India aims to have 250 GW of solar and 100 GW of wind by 2030, with a 40% overall renewable power target, large hydro excluded. Interestingly, there are reports that India is cutting back on coal use since renewables are now seen as cheaper (Russell 2019), resulting in falling emissions (Myllyvirta 2019). So India could be a major player in renewables despite its parallel dalliance with nuclear power.

Brazil has also done quite well. It gets 87% of its power from renewables, mostly hydro, but although Russia has more than 50 GW of hydro, supplying 16% of its power, it is still at the starting gate on other renewables: it aims to get only 4.5% of its power from non-hydro renewables by 2020.

Its main current focus is fossil fuel, including export, and its main future focus is nuclear, including export of its nuclear technology.

Japan used to be seen as a major technology player, and it still is in many areas but not so much in renewables. It had focused on nuclear. However, after the Fukushima nuclear accident in 2011, it had to close all its nuclear plants and, although some are restarting, it is trying to catch up on renewables (Elliott 2013b). It aims to expand renewables so that they supply 22–24% of its power by 2030. That may seem a rather low target, given what is happening elsewhere, but, as noted in chapter 6, in addition to being late to the game, Japan, with its high population density, has major land constraints and so little room for onshore wind or large biomass or solar farms. It might therefore be seen as one of the worst-case locations for renewables and is still using significant amounts of (imported) fossil fuel and is keen on CCS. Nevertheless, it is also pushing ahead with rooftop PV (over 55 GW so far) and with floating PV on reservoirs, as well as offshore wind, including floating projects. It has also taken energy saving seriously, this being an urgent requirement to avoid blackouts post Fukushima.

After Fukushima, and then a national nuclear safety scandal, South Korea decided to move away from nuclear and aims to get 20% of its electricity from renewables by 2030. Taiwan has also backed off from nuclear and intends to phase it out and get 20% of its power from renewables by 2025, although, as in South Korea, the phase-out is contentious and may be revised or delayed. Meantime, Vietnam has decided not to go ahead with a Russian-backed nuclear project and is focusing on renewables. It has signed up to an ambitious 100% by 2050 renewable power target, an aspiration shared with the Philippines and Papua New Guinea.

The potential for renewables in these and other south-east Asian countries is very large. Bent Sørensen has produced near-100% scenarios for Japan and South Korea, as well as for India, with full grid balancing (Sørensen 2014), and further national scenarios for the region have been developed as part of '100% renewables' studies by the LUT/EWG and Jacobson's team, along with others.

Some of these studies cover Australia and New Zealand. Neither of these countries has nuclear and both have enviable renewable potentials, but Australia has been relatively slow to develop them and, being a major coal exporter, has been ambivalent on climate change issues, despite being in the front line of climate impacts. It has a renewable power target of 23% by 2020. By contrast, New Zealand aims to get 90% of its power from renewables by 2025 and to transition to 100% renewables by 2035, and it hopes to become completely 'zero carbon' by 2050.

There are many other players in the renewable energy field around the world, and I have reviewed the overall state of play around the world in detail elsewhere (Elliott 2019a). The general picture is one of quite rapid growth in many areas but not in all. For example, most countries in the Middle East are still looking to oil if they have it, although, as I indicated earlier (chapter 6), that may be changing, with the take-up of solar.

One way to try to identify which way things might go in the future is to follow the money. As I noted earlier in this chapter, funding levels for renewables have been cut back in the EU and also in China, but cumulative capacity growth continues globally and regionally. Driving that, the pattern of new investment is interesting: it is increasingly skewed towards China and the developing and emerging economies and away from the already developed economies, with China's huge expansion (although reduced of late) clearly tipping the balance between what might be seen as the 'old' and the 'new' worlds (see Box 7.4).

Box 7.4 Global investment patterns and sources

Investment in clean energy technology has boomed in recent years globally, reaching more than $300 billion per annum, although, given China's slowdown, overall growth fell back in 2018 (BNEF 2019). In terms of total financing of renewable energy (excluding large hydro) according to REN21, China accounted for 32% of the global total in 2018 (down from 45% in 2017), followed by Europe (21% in 2018), the United States (17%) and Asia-Oceania (excluding China and India) 15%. Smaller shares were seen in India (5%), the Middle East and Africa (5%), and the Americas (3%), excluding the United States and Brazil (REN21 2019).

These investments typically come from the finance sector, banks and in some cases sovereign wealth funds, with governments and state-owned utilities sometimes playing a role, as do international funding agencies. For example, the World Bank and various UN and EU funding programmes invest in some development programmes.

However, in terms of technology, the private sector tends to dominate. Power utilities and major energy companies, including some oil companies, sometimes support renewable development and may provide investment capital for new projects. There are also many engineering companies active in the field around the world, ranging from giants like Siemens based in Germany and GE based in the United States to smaller more specialist ones like Orsted in Denmark and Sinovel in China. They may support new technology development from their own sources, but usually borrow investment capital from the finance sector for large projects.

REN21 notes that 'private sector investment and procurement decisions are playing a key role in driving renewable energy deployment. As of early 2017, 48% of the US-based Fortune 500 companies had targets for emissions reduction, energy efficiency or renewable energy or combinations thereof.' In terms of corporate sourcing of renewable energy, interestingly, REN21 says that 'the IT sector continues to purchase by far the largest amounts of renewable energy. The three largest corporate buyers in all sectors at the end of 2017 were US-based Google, Amazon Web Services and Microsoft, all of which source renewables around the world', usually via power purchase agreements (REN21 2019).

It is hard to predict what might happen next. It will depend in part on the relative strength of the drivers of change, including pressure from NGOs, environmentally concerned consumers and progressive shareholders, and resistance to change. There is no shortage of the latter, with fossil fuel interests often playing a role. Coal use is still widespread in many parts of the world and even expanding in some areas (Carbon Brief 2019), but it is being widely challenged (Shearer 2019; Shearer et al. 2018). So too is oil use. The oil companies will no doubt want to avoid major changes as long as they can, but some of them are making adjustments and diversifying their portfolios to some extent, including by moving into solar and wind (Shojaeddini et al. 2019).

That may be relatively marginal so far, and gas use is expanding in some areas, including shale gas use, with some seeing gas as a slightly cleaner interim option. However, in addition to problems with transmission leaks and fugitive emissions (methane is a powerful greenhouse gas), burning it still generates CO_2, and even if that might be partially reduced with CCS, relying on fossil gas is not a long-term solution.

It may be hard for some countries, but it seems clear that we all have to stop burning all fossil fuels. That view is now spreading, with concerns about climate change mounting worldwide, amongst young people especially, but also amongst companies (Murray 2019). That is not surprising since, as has been argued above, there are direct economic incentives for making this change at all levels. For example, in terms of renewable electricity generation and use, in addition to local-level cost-saving advantages for prosumer and community projects, there are national-level commercial, operational and trade incentives driving interest in the use of renewables. Globally, 185 companies have already signed up to 100% renewable power targets (RE100 2019) and, as noted in Box 7.4, there are welcome signs of corporate change, with key companies getting into, benefiting from and creating a booming green energy market.

Longer-term, as noted in chapter 6, global trade in green gases and liquids, as well as regional and perhaps even global trade in green power, could become widespread. That may be some way off, but the spread of renewables is underway with, by 2018, more than 50 countries having adopted '100% by 2050' renewable electricity targets (REN21 2018). So the world is changing, country by country. That said, outside the EU, few countries have renewable targets other than for *electricity*. Heat and transport energy is mostly not yet covered and, worryingly, energy efficiency rarely gets taken seriously. So, as the review above illustrates, although progress is being made around the world, there is still a long way to go.

Prospects for the future

In a paper produced for the G20 summit in Japan in 2019, the IEA says that:

> Even if wind and solar PV deployment could be accelerated, other low-carbon technologies like dispatchable renewables, nuclear power and CCUS also need to be expanded at massive scale to decarbonise the power sector. The level of additional renewable generation sources required to achieve the Sustainable Development Scenario is already extremely high. Expanding the level even more to make up for the lack of growth or decline in nuclear power or CCUS implies enormous challenges in terms of not only additional costs but also land availability and local acceptance. (IEA 2019f)

Plainly, the IEA did not think it could be achieved just using renewables, at least not easily or quickly. IRENA's paper to the G20 Summit was more optimistic. It reiterated its view that 'renewables and energy efficiency, boosted by electrification, can provide 90% of the necessary reductions in energy-related carbon emissions to limit the global rise in temperature to well below 2°C by 2050. Indeed, renewables and electrification alone would provide 75% of the reductions needed' (IRENA 2019d).

As I have indicated, others think even more could be achieved, all the way to 100% of all energy, if proper attention were paid to the demand side. In addition to their global scenarios, the LUT/EWG team has developed ambitious '100% renewables' scenarios for most parts of the world, as have Jacobson's Stanford team. They go a bit beyond some of the earlier NGO studies of what might be possible, although these too saw the renewable energy share expanding dramatically. For example, a WWF/TERI study of India suggested that it might be able to supply 90% of all its *energy* from renewables by 2050 (WWF 2013). A bit more cautiously, a World Wide Fund for Nature (WWF) study claimed that China might be able to get 80% of its *electricity* from renewables by 2050 (WWF 2014). Greenpeace have also produced a series of 'high renewables' national/regional studies based on its '100% renewables by 2050' Energy [R]evolution scenario (Greenpeace 2015), while the Global

Energy Network Institute (GENI) has produced a range of '100% renewables' national studies (GENI 2016).

It remains to be seen if expansion on the scale envisioned by academics and green NGOs is really possible globally. The evidence from this chapter is mixed. There are some inspiring changes underway in some countries; as mentioned above, more than 50 have signed up to get to 100% renewable electricity by 2050. However, that is not the same as 100% of *all energy*. So far, only a handful of countries have signed up formally to 'net zero carbon by 2050' targets or similar. Moreover, 2050 is some way off so making long-term commitments is a relatively easy gesture. Living up to these targets may be another matter, especially given the pressure of population growth in many developing countries. For example, the population in Africa is expected to at least double by 2050. In addition, some developing countries are effectively locked into oil exports as a key part of their economy, Nigeria for example. 'Big oil' is still very powerful, and parts of the oil lobby can certainly at times sound very bullish, at times arguably counterproductive in its own terms (Watts 2019). More generally, for developing countries facing massive social and economic problems, climate change may seem far off and less important, even irrelevant.

Clearly, there are social economic and political issues to face in the developing world as elsewhere. Nevertheless, in terms of technology, while not enough attention is being paid to energy efficiency and demand reduction, progress is being made in most places on renewable supply, driven by falling costs, along with concerns about climate change. That on its own may not be enough to bring about change on the scale needed, but it may help to start things off. In the past, renewables were sometimes seen by hard-pressed developing countries as expensive and mostly irrelevant to their growing needs. Now, with lower costs, they are beginning to move centre stage and are increasingly being seen as vital for their development (Elliott and Cook 2018).

While it may be true that the overall energy transition and the renewables expansion programme is taking time and may be too slow, this is due to a mix of technological factors and wider social, economic and political factors. The optimistic view is that the technology is mostly available, or can be made so, but we need the political will and social and economic change to create a viable way forward.

The pessimistic view is that none of that is going to be easy or even that most of it is irrelevant. For example, the Chinese renewables programme is sometimes portrayed as a side issue compared to its continuing use of coal, and the scale of its emissions is sometimes used to suggest that it is hardly worth anyone else trying to cut theirs. Continued reliance on fossil fuel is certainly a problem across most of Asia, in China in particular. Japan is also still struggling with the results of the collapse of its nuclear programme and is importing gas and coal to compensate, as well as oil for cars. Meanwhile Australia, the Gulf states, Russia and more latterly the United States, amongst others, seem happy enough to export fossil fuels to keep the show on the road.

So there are problems still to resolve and there is never any shortage of gainsayers keen to highlight any failures or slowdowns. For example, those wedded to the existing power system, including some with contrarian views about climate change, are often disparaging about the 'flagship' German *Energiewende* energy transition programme, depicting it as flawed and failing. The UK Global Warming Policy Foundation has been amongst those producing critical commentary (Vahrenholt 2017).

Despite being generally popular with the German public (Wehrmann and Wettengel 2019), as I have indicated, the German *Energiewende* transition programme does have issues and is slowing, mainly due to policy pressures, and that is a cause for concern. However, not to all. Underlying the hostility of at least some critics may be resentment of Germany's nuclear phase-out and despair that leading engineering companies like Siemens have backed renewables instead, a shift in view also adopted more or less enthusiastically by major German power companies RWE and E.ON. Even worse, the heresy has spread, E.ON famously withdrawing from the Horizon nuclear project in the United Kingdom, saying, 'we have come to the conclusion that investments in renewable energies, decentralized generation and energy efficiency are more attractive, both for us and for our British customers' (Teyssen 2012).

Battles over corporate and government policy will no doubt continue but the bottom line, shared by nearly all parties, is that if we are to avoid the worst impacts of climate change, there will be a need for technological and policy change around the world. The improved economics of renewables may help, but governments also need to take action and to promote, and aid, rapid adoption of renewables, as well as energy saving. Some even talk of going onto a war footing. Certainly, decisive action is urgently needed. It is easy to be depressed by the current, often relatively slow progress in many countries but, as this chapter should have illustrated, there are also grounds for hope that the huge global potential of renewables will be exploited, helping to avoid a global climate breakdown.

8
Conclusions

How much can we expect from renewables as we try to move to a sustainable energy future? This chapter rounds off the discussion by looking at prospects for, and costs of, an accelerated renewable energy expansion programme.

Can it be done?

It is sometimes argued that renewable energy is needed but will not be enough to deal with all our environmental problems, some claiming that the pace of change is not sufficient to avoid major climate change impacts. So far, the national and global policy emphases have often been on emissions-reduction targets, sometimes set far into the future, but now some say we are faced with an urgent need for much more rapid progress.

Some analysts are optimistic. For example, IRENA envisages a future in which renewables, aided by demand reduction, take over most of the energy supply and, as I have indicated, others have claimed that they could take over it all (Jacobson et al. 2017; Ram et al. 2019). They are all certainly more optimistic than the recent report from the World Economic Forum, which gloomily says that, although some progress has been made in some countries, the world's energy systems overall have become less affordable and are no more environmentally sustainable than they were five years ago, with in some cases emissions still rising (May 2019).

Spencer Dale, group chief economist at BP, has put the situation starkly: 'Despite extraordinary growth in renewables in recent years, and the huge policy efforts to encourage a shift away from coal into cleaner, lower carbon fuels, there has been almost no improvement in the power sector fuel mix over the past 20 years. The share of coal in the power sector in 1998 was 38% – exactly the same as in 2017' (BP 2018a).

To the extent that this is true (and actually it may be changing, as coal phase-outs spread globally), it is because energy demand has risen, with fossil fuel use expanding to meet it, so that the fossil proportion in the mix has not fallen, even though renewables have expanded. However, it is also the case that, as REN21's 2019 Renewable Energy Status Report notes, although renewable growth had reduced from the results it quoted in its 2018 report, modern renewables have still expanded their share in final energy consumption by an average of 4.5% over the last ten years, whereas global energy demand had only risen by 1.5% (REN21 2019). So, arguably, they are still doing quite well – indeed, in absolute terms very well, with wind and solar power output growing 'at an annual average of 20.8% and 50.2%, respectively, over the past decade' (Rapier 2019).

Nevertheless, renewables will have to try much harder if they are to squeeze fossil fuels out and if demand stays high. The United Kingdom seems likely to be able to move quite rapidly in that direction. It currently gets nearly 40% of its power from renewables, and the ambitious 'community renewables' scenario outlined by the National Grid company, wind and solar dominate power supply, in all renewables generating more than 75% of UK electricity by 2030 and hydrogen being used for some heating and transport. Energy demand is dramatically reduced and local generation is emphasized, with biomass being used for some local heat networks and with less than 6 GW of nuclear. So it is a mostly non-fossil future in terms of electricity supply and heading that way for heat and other energy uses (NG 2018). The UK Labour Party commissioned a report on the options that outlined a somewhat similar package. It looked to cutting power-related emissions by 77% by 2030, with renewables ramped up rapidly to 137 GW and demand cut by 20% for heat and 11% for electricity, although it still retained 9 GW of nuclear (Labour 2019b).

More radical approaches, with more attention being paid to energy saving and renewables accelerated further, could eliminate the need for nuclear and squeeze fossil fuel out of most of the rest of the economy. Many NGO and academic studies over the years have suggested that getting to or near 100% renewables in most sectors was viable for the United Kingdom by 2050, or even earlier (CAT 2013; FoE 2017; Pugwash 2013; RSPB 2016; RTP 2015; WWF 2011). A variant of this type of programme, with no new nuclear power, has recently been proposed by the UK Liberal Democrat Party, with renewables supplying 80% of electricity by 2030 and setting a target of 'net zero carbon' by 2045 'at the latest' (Lib Dems 2019). That may be pushing it for all sectors, but scenarios with renewables supplying 100% of all energy by 2050 have been outlined in the UK/ Ireland regional subsets of the global models produced by LUT/EWG and by Jacobson's team.

As was indicated in the last chapter, similar scenarios have been produced for many other countries, including by NGOs like Greenpeace (Greenpeace 2015), WWF (WWF 2013, 2014) and GENI (GENI 2016). Although some of them are just for electricity, some of the more recent academic studies, like those by LUT/EWG and Jacobson's team, go beyond that to 100% of all energy by 2050. That would require very rapid expansion of renewables, in some cases from quite low current levels, along with technical measures to ensure that energy was used more efficiently.

Is that credible? Expansion on the scale necessary may be possible in some countries, most obviously those that have already achieved high renewable power shares, and getting to near 100% of power does seem credible for many of them. However, achieving 100% of all energy by 2050 seems likely to be very hard in many cases, and quite hard even for the 'advanced' countries. The United Kingdom is blessed with excellent renewable energy resources (offshore especially) and might be able to achieve something near that, assuming new policies; Denmark too, but Germany's current renewables target is only for 60% of energy by 2050. They, and others, may be able to do better than that, with upgraded policies, and there are now strong pressures for more rapid expansion, some 25 countries now aiming for 'net zero carbon' by 2050 or similar targets.

Nevertheless, there is still a way to go globally, and there are those who doubt that technological and institutional change on the scale necessary can be made rapidly (Lyman 2019). In terms of technology, changes like this inevitably involve slow processes; in some cases it takes decades to move from invention to wide use (Gautier 2019). In chapter 2, I mentioned that counters to this view had been offered, arguing that more rapid deployment of renewables could be made, given the right support structure.

However, the alternative, negative view does resurface regularly in various guises. For example, in a report calling for nuclear expansion, the IEA noted that in the past 20 years wind and solar PV had increased by about 580 GW in advanced economies but, if we are to meet the climate challenge without nuclear, 'over the next 20 years, nearly five times that amount would need to be added'. It states that 'such a drastic increase in renewable power generation would create serious challenges in integrating the new sources into the broader energy system' (IEA 2019e).

Subsequently, in its 'Renewables 2019' market review, the IEA did predict that the world's total renewable-based power capacity would grow by 50% between 2019 and 2024, an increase of 1,200 GW in five years, so maybe even more rapid increases are not out of the question in the years ahead (IEA 2019g). The IEA's predictions of renewable expansion rates do seem to have been consistently low over the years (Evans 2019b).

A similarly rather pessimistic view on renewables was, a little surprisingly, also offered in a Bloomberg New Energy Finance post by Michael Liebreich. He was scathing about old nuclear but said we will need new nuclear since renewables cannot expand fast enough globally, at least by 2030:

> If your plan to deliver a 20% or 45% emission reduction in the electrical sector – targeting 2 degrees Celsius or 1.5 degrees Celsius

of warming respectively – is via wind and solar alone, assuming some moderate level of economic growth, you would have to add two to four times as much capacity in the next decade as has been added in total in the last two decades. BNEF's recently released New Energy Outlook 2019 shows that, while we could hit the lower end of that range, it is highly unlikely we will hit the higher end of the range on the current trajectory. (Liebreich 2019)

Liebreich and BNEF are more usually known for charting the rapid expansion of renewables, for example, talking of a 90% power share by 2040 in the EU, 80% from wind and solar (Romm 2019), so the rendition above, albeit only for 2030, may seem rather conservative. Liebreich's view on the prospects for nuclear also seems to contrast with another post from BNEF, which sees nuclear as being irrelevant unless it gets cheaper (Tirone 2019).

However, it may yet do, and the 90/80% BNEF figures for renewables quoted above are the share just for the *EU*, whereas BNEF only see renewables in all getting to 50% of power *globally* by 2050. That may of course be pessimistic. Certainly, as I have indicated, there are more radical potential expansion trajectories, ramping up to very much higher levels by 2050, especially given the possibility that demand could be radically constrained.

Liebreich admits that energy efficiency could reduce demand growth and energy intensity, possibly even by an extra 25% by 2030. Actually, the EU target is a 32.5% demand reduction by 2030. But leaving that aside for a moment, Liebreich said that, to meet the heat and transport demand, as well as power demand, would mean 'building an additional 10 to 15 times current installed capacity of wind and solar', the higher level being required for the case when less energy saving was possible.

While '10–15 times' would certainly be stretching it for wind and PV, the potential is certainly there. IRENA's relatively conservative estimate is that there could be around 6 TW of wind capacity in use globally by 2050, along with 8.5 TW of PV solar (IRENA 2019e) and, as I indicated in chapter 3, others have put the 2050 potentials very much higher, at many tens of TW each. Moreover, we can also use electricity from a range of other renewables (not just wind and PV) including wave, tidal, solar-thermal CSP, geothermal and, possibly, some hydro and biomass/biogas sources. Their output, along with that from wind and PV, can be used directly to meet power needs and also for some vehicles and some heating. In addition, we can use solar, biomass wastes and geothermal directly for heating, and green vehicle synfuels for some transport. It does not all have to be done by wind and PV or just by green *electricity*.

It also does not all have to be done just by adding more *supply* capacity of whatever sort: *demand* can be reduced further, hopefully by much more than the 25% top level that Liebreich envisages. It may take time. For example, Germany is aiming for a 50% primary energy cut by 2050. However, the fact that electricity demand is falling in some countries (see chapter 6) suggests that an easier transition may be possible.

Certainly, as I argued in chapter 2 and elsewhere in this book, reducing energy demand will be very important if we are to meet the climate challenge. It can be attempted both via technical means and by social adjustments. The technical measures are relatively straightforward, but the social changes may be more difficult, given that significant lifestyle changes could be needed if the reduction in energy use was to be substantial. Nevertheless, with climate impacts likely to become ever more visible, it may not be out of the question to win support for that and also for the rapid expansion of renewables.

However, it may be that, for whatever reason, the necessary technical and social changes on the supply and demand sides will not prove to be possible or that the rate of change will turn out to be insufficient to avoid the full and growing costs of climate change. In that case, then, as was indicated in chapter 5, since some say that growth in energy and material use is fuelled by economic growth, we would need to slow or even *halt* economic growth – a much more serious move, as Liebreich would no doubt agree (Liebreich 2018).

The growth debate

Given the rapid fall in renewable energy costs, Professor Hinrichs-Rahlwes of the German Renewable Energy

Federation told the 2018 World Renewable Energy Congress in London that 'we no longer have to choose whether protecting the environment and limiting global warming is more important than growing the economy or vice versa' (Hinrichs-Rahlwes 2019).

However, some 'deep green' environmentalists argue that, on a planet with finite resources and carrying capacity, even given clever low-cost technology, we cannot have both economic growth and environmental sustainability. The radical greens challenge the 'ecomodernist' view that, with advanced energy technology, economic growth can be decoupled from environmental limits (Heinberg et al. 2019). Instead, while not opposing the 'stable-state' use of green energy technology, the 'deep green' critics say that, rather than seeking what they see as an illusory high-technology cornucopian future, we need a 'de-growth' approach, and a new form of economic interaction (Hornborg 2019).

The debate on growth can get quite complex, with a recent input from Professor Vaclav Smil, the celebrated energy analyst from the University of Manitoba in Winnipeg, taking a very wide approach. He has little time for techno-optimism or the decoupling theory. Based on a very wide-ranging review of growth in natural systems as well as human systems, he concludes that economic growth must end (Smil 2019). That is a quite challenging view, akin to the 'deep green' view that the only hope is radical social change and a move away from growth.

The pessimistic ecological view can certainly be bleak. Nature combines growth with decay in a balanced way. We do not: so far, our technology has mainly been based on ever-expanding consumption of finite resources and dumping any waste without too much care. It may of course be possible to change, but views differ on what that would entail in terms of energy systems and much else. Some say that we will need both social and technical change, and that, at the very least, significant social changes will be needed to make technical change viable. Put more positively, and focusing just on energy, social changes leading to less energy use would make it easier for renewables to meet reduced energy demand.

However, some say that energy is not the only issue and that, in any case, we need less emphasis on technology and much more emphasis on social and economic change. For example, some look to the localization of economies and a shift to sustainable consumption within the context of a low- or zero-growth economy. More specifically, focusing on energy, some say that, rather than just looking to technical efficiency, as Lovins suggests (Lovins 2018), we should move to a 'sufficiency' approach, reducing our demands to only what is sustainable (Parrique et al. 2019).

That of course may require some major social changes. For example, some say that the energy demand reductions needed cannot be achieved without severe constraints on energy use, implying rationing either by price or by some other mechanism, possibly coupled with or leading to a halt to economic growth. Prescriptions like this open up a range of social equity issues. In the bleakest visions of the future, these changes and worse will be forced on us if we do not act in time to avoid environmental and ecosystem collapse.

Most of these views on ecosystem collapse, social change and growth contrast strongly with the currently widespread view in official circles that economic growth and sustainable development can be compatible (the 'decoupling' view) and that social change may not have to be too severe. Certainly, a degree of decoupling does seem to have occurred between energy use and economic activity in some countries. That does ignore the trend to 'offshoring', shifting high-emission industries to other countries while importing their products. Nevertheless, it has been claimed that *within* the UK economy, decoupling of energy demand from economic activity 'has contributed more to carbon emissions reduction than the combined effects of the UK's programmes in nuclear, renewable and gas-fired power generation' (Eyre and Killip 2019).

However, although there is a huge potential for avoiding energy waste, there may be limits to how much demand (and emissions) can be continually reduced by improved efficiency and changing energy use patterns, if economic growth (and offshoring) is not constrained. It may not be possible to continually accelerate the rate at which you can squeeze demand (and carbon) out of the system to compensate for a system that is expanding rapidly. Some nevertheless say that emissions can be cut by the use of renewables so that, if we want, we can continue with economic and energy use growth.

Others say that there will be limits to how widely renewables can be adopted, and there may also be non-energy limits to continued economic growth (see chapter 5). So the argument over the viability of 'sustainable growth' continues.

At the same time, there are also limits to the alternative approach, halting economic growth to cut energy demand and eco-impacts. It would have many social equity implications. It would also be fought hard by those with vested industrial and commercial interests in growth. That on its own is no reason to avoid change, but the 'de-growth' view has also been challenged from the political left, with claims that it would lead to eco-austerity: 'the immediate effect of any global GDP contraction would be huge job losses and declining living standards for working people and the poor.' By contrast, it was claimed that a well-designed Green New Deal approach could avoid that, while cutting emissions, by investing around 1.5% of GDP p.a. in clean energy (Pollin 2018).

Perhaps what both sides in this debate can agree on is the benefits of growth in *renewables*, at least for a while, as part of the transition to a sustainable future. There will still be room for debate about what sort of renewables to use and how they are developed. For example, 'deep greens' may not want to see renewable growth being used to enable the continued expansion of consumer society and rapacious economic growth. Looking longer term, there will also be debates over what the end-state should be. Some greens look to a stable low- or zero-growth future, when a fully renewable system has been established, needing only maintenance and the occasional replacement of older systems with new, upgraded, more efficient systems.

However, although some replacement/upgrade projects for old green energy systems are already underway, we are some way off having to face that issue on a grand scale, or for that matter the growth issue, at least in terms of renewables: they still need to grow. For the moment, the main issue is getting started fully on the transition, with a key debate being the relative importance of technical and social change in the interim. That of course also applies to the longer term, defining what sort of society we want live in.

On the social change side, in both the short and long term, key questions are how much social change will be needed and should be aimed for, and also how much of whatever change is needed can be achieved in reality. That essentially political debate continues, and I will be coming back briefly to it later on and also to the issue of whether we need a shift from quantitative to qualitative growth. However, my main focus in this book has been on the technology side, with the case for that being a major focus in the transition and perhaps also in whatever society subsequently emerges arguably being quite strong.

The green technology option

As I have argued, although the social change aspects are important, and there are social and political constraints on what might be achieved technologically, there may be room for some techno-optimism, at least in terms of dealing with the emissions that lead to climate change.

Certainly, the authors of most of the energy scenarios I have looked at that have very high renewables proportions avoid prescribing major social change or zero growth. They can do this with some conviction because they argue that technological systems change, along with some social change, can deliver the necessary carbon reductions. Some of this, it is argued, would involve energy demand reduction, through the more efficient use of energy via a range of end-use technical fixes, possibly requiring minor changes in consumer behaviour. However, the big savings would come from overall system change.

A key point is that switching from the use of fossil and nuclear fuels in low-efficiency steam-raising systems to a system using electricity generated directly from renewable energy flows, like wind and solar, will lead to large reductions in primary energy use. It is not just a case of avoiding the use of fossil or fissile energy sources but also of avoiding the energy wasted in the conversion of these fuels to electricity. It is worth unpacking that issue a little.

It is matter of the basic energy-conversion thermodynamics involved. Old coal-fired plants typically generated power at around 30–35% energy-conversion efficiency: most of the energy in the fuel used (the so-called primary energy) was wasted as heat was ejected into the environment (see Box 1.3). Combined heat and power (CHP) 'co-generation' plants make use of that waste heat and so raise their overall energy-conversion efficiency (for heat and power) to maybe 70–80%. A little less effectively, combined cycle gas turbines (CCGT) use the hot gases from the first gas-turbine generation stage to produce steam for a second power generation phase, raising overall efficiency to around 50%. In both cases, to differing extents for fossil-fired CHP and CCGT, less fuel is needed to get the same output, and emissions/kWh are reduced. However, fuel is still used, and the same is true for nuclear plants, which still waste much of the energy released from their fuel unless they are run in CHP mode.

By contrast, renewable energy systems, like wind turbines and solar cells, do not need any fuel to run, and that is also true for solar heat-based systems, as well as tidal, wave, geothermal and hydro power, so in principle their *primary* direct energy use is *zero*. As all energy systems, they will need energy for their construction but, as argued in chapter 5, in time that can come from renewable sources. So, in theory, we can move to a zero-carbon system in generation terms, with renewable power being used in all sectors. That would lead to large overall energy savings, not just in primary energy, but also in efficient end-use terms. As noted earlier, Jacobson's latest scenario foresees global end-use energy demand falling by 57.9% by 2050, due to (primary) fuel substitution, end-use electrification and energy-efficiency upgrades, with renewables then meeting the much reduced residual demand in all sectors (Jacobson 2019a).

As I noted in Box 3.1, most oil companies do not see it like that; demand for energy and electricity will rise and renewables will not be able to meet it. That picture was reinforced by the most recent comparative study of scenarios from oil companies, as well as from the IEA, EIA and so on, in a Global Energy Outlook (GEO) produced by US-based Resources for the Future (Newell, Raimi and Aldana 2019). The scenarios it looked at nearly all depict energy use expanding up to 2040, much as was found in the earlier comparative study by the World Energy Council (see chapter 3). In the GEO review, renewables reached at best a 31% primary energy contribution by 2040, and energy demand expanded in nearly all scenarios and in all end-use sectors. In some cases, recourse was made to carbon capture and negative emissions technologies to compensate for the continued use of fossil fuels, but emissions still grow.

Clearly, there are some divergences in views, some of them due to methodological differences, concerning whether use is made of primary energy or delivered (final) energy in the assessments (see Box 1.3), and some concerning the levels of end-use electrification assumed. For example, Sverre Alvik, director of the Energy Transition Outlook programme run by global engineering consultants DNV GL, says, 'the strong electrification of all sectors, but most of all in the transport sector, and the "high renewables" share, are two distinct differences from DNV GL's forecast and many scenarios from other forecasters, and this is the main reason for DNV GL seeing world energy peak, while most other forecasters do not' (Hill 2018).

DNV GL's scenario has renewables supplying about half of global energy by 2050 with, in its 2019 Energy Transition review, renewables making up 80% of the *electricity* mix by 2050 (Merchant 2019). So it is high but not up in the '100% league' with LUT/EWG and Jacobson and colleagues. These two groups may be the leading proponents of the '100% renewables' view, but they are not alone. As I noted earlier, there are in all around 42 academic studies from 12 different research groups worldwide, suggesting that energy for electricity, transportation, heating/cooling and industry could be supplied reliably with 100% or near-100% renewable energy (Stanford 2019). In addition, there are the global and national studies mentioned above, carried out by NGOs. However, it is notable that none of these more optimistic academic or NGO studies feature in the GEO review, nor in the earlier WEC review of scenarios (see chapter 3).

Interestingly, nearly all of the NGO/academic studies see wind and PV leading, but some of their scenarios also include significant amounts of non-electrical renewables for heating and transport, including biomass, with CHP and local energy generation often promoted more. However, some scenarios exclude or downplay biomass, as I explored in chapter 3, where I indicated that there have also been environmental concerns about large hydro. So there is quite a range of scenario mixes, with variable amounts of energy saving also being assumed.

It has to be said that, although they may test general feasibility, none of the scenarios from any group can prove that 'high renewables' shares will (or won't) work. That will only be found out by trying. As can be seen, the weak points are likely to be transport and heat supply, both hard to model, with a multiplicity of possible lines of development and end uses but, so far, with some being undeveloped or even hypothetical.

There are also other uncertainties. We have to be prepared for impacts due to climate change, which may modify the renewable resource, for example altering wind and solar regimes. As I have noted, climate change is already having an impact on hydro, but one recent study found that the wind and wave resource was actually improving (Young and Ribal 2019). It is hard to predict trends like this with certainty and there may well be unexpected problems ahead.

Nevertheless, on the basis of the studies reviewed in this book, some reasonably clear technical conclusions can be drawn. There may be some laggards and some who will find it hard. However, around 100% of renewable electricity supply seems to be within our reach globally, possibly by 2050, maybe earlier, although 100% of all energy may take longer, particularly in relation to transport. I suspect this is where lifestyle changes and changed consumer behaviour will have to play more of a role, although much will depend on how the new technologies develop and, crucially, on their cost.

How much will it cost?

Certainly, a key issue for the energy transition, shaping whether, or how widely and rapidly, it might happen, is the cost. At various points in this text, I have reported on claims that, long term, the overall cost of running a sustainable energy system could be lower than that of the current system. It would avoid the high and rising social, environmental and economic costs of fossil fuel use, and the high costs and risks of nuclear technology (including long-term active waste management) and, by better matching varying supply and demand, could raise energy use efficiency, reduce peaks and cut consumer costs. For example, some of the smart grid-balancing options would reduce energy waste while supergrid links would create new opportunities for the potentially lucrative power export trade.

The economics of a fully developed and balanced green energy system do look quite promising, at least in theory, and they could improve further in time. For example, although there may be transitional costs, as I noted in Box 2.1, once renewables expand beyond supplying 100% of power demand most of the time, further balancing costs are avoided and the extra power and surpluses can be used for other end uses, earning income to offset the extra cost.

Nevertheless, there will be costs in establishing the new system. In chapter 5, I noted the claims from the UK government and the European Commission that the transition costs would be offset by the savings from the programme so that the net costs to consumers should not rise. That might not always be the way it turns out in practice (see Box 2.2) but, in principle, it should be possible to limit any extra costs by careful design of the programme. Crucially, the falling costs of renewables should help; so should P2G conversion if it turns out to be as viable as some hope, enabling easier and beneficially productive balancing.

However, that may all take time. For now, as I noted in chapter 1, given that there have been backlashes against current cost rises, the big political issue may be whether the falling cost of renewables and also of storage/balancing can ensure that the cost of rapid expansion will not provoke consumer resistance. That will depend to some extent on how rapidly and how far the expansion is pushed ahead. There is certainly an environmental case for going faster and further than in current plans and, as I have hopefully demonstrated, the technological case for that being possible is strong. It can also be argued that there is a wider economic case for that too so as to avoid the huge future costs of climate change, although going faster and further may impose higher transitional costs.

In chapter 5, I argued that, if that was the case, it would be vital to ensure that any extra costs incurred were shared fairly so as to avoid conflict and resistance. What is still unclear is whether this problem can be avoided or limited by the continued fall in renewable energy and balancing system cost, as well as cost savings from energy efficiency. The fall in cost for renewables and some storage has been dramatic and seems to be ongoing. There are also claims that the cost of energy saving will fall. Hopefully, that will all continue, with system costs falling along the lines that some (see Box 2.1) suggest will be possible.

To some extent, then, we are in the hands of the technologists and those working to get costs down. There are ways in which this process can be aided, in addition to relying on conventional market competition. For example, appropriate and consistent levels of financial support can be provided for each stage of the innovation cycle, from R&D and project demonstration and on to early market deployment. Governments have usually been happy to support R&D, but most of the cost-reduction gains often occur in the latter market take-up phase (Kazaglis et al. 2019).

Guaranteed price feed-in tariffs (FiTs), as used initially across much of Europe, proved to be very effective at getting capacity established and expanding the market, and that market-enablement process helped get prices down rapidly but at a cost to consumers. In effect, they acted as pioneers, funding the market acceleration, which then led to lower costs, enabling continued expansion. Nevertheless, this continuing process did push up pass-through consumer prices and that led to a reaction. FiTs have now been replaced by competitive contract auctions, which may avoid the cost pass-through to consumers and force prices down further but may limit the rate of deployment, with only the lowest-cost projects winning. Some say that, if you want fast and widespread deployment, you have to provide extra support: someone has to pay. However, equally, it is clear that, if prices can be brought down, then over time market volume is likely to build, in theory lowering prices further.

Learning curve projections, based on past cost and performance trends for new technologies as they pass through the innovation/market uptake cycle, do indicate that cost falls will continue for all the new energy technologies (wind and PV especially) as their markets build. DNV GL says it expects the 'learning rate of 18% for PV and 16% for wind to continue throughout the forecast period', i.e. up to 2050. However, learning curve analysis is not always a reliable or accurate predictive tool since there can be discontinuities, breakthroughs and collapses.

It may therefore not be possible to construct realistic cost curves to show the balance of falling unit costs versus rising overall expenditure, not least since we are also talking about hypothetical programmes running into the future and set against unknown counterfactual opportunity costs. That may mean that is a judgement call, dependent in the end on broad assessment of the risks and the potential for cost reduction. From what I can see, though, we are in a race to get costs down as fast as possible so as to avoid having to face much higher costs if we fail to expand renewables and efficiency fast enough and get hit by massive and ever-rising climate-impact costs.

In the UK context, the Treasury has calculated that the costs of the now-agreed move to 'net zero emissions' by 2050 will be around a total of £1 trillion. That is 40% higher than the Committee on Climate Change's estimate for the cost of getting to net zero by 2050. However, the former leader of the Labour Party and co-chair of IPPR's Environmental Justice Commission, Ed Miliband, thought it was a good deal anyway since it was only 1–2% of GDP, whereas the cost of climate change could be vastly more (Grundy 2019). Estimates vary but the classic Stern review has suggested that the overall impact costs of climate change, if unabated, could be the equivalent of losing up to 20% of UK GDP each year (Stern 2007).

On that basis, a £1 trillion package for the United Kingdom, and no doubt similar sums pro rata elsewhere, might be seen as an expensive but necessary insurance policy or, more positively, as a wise investment for securing a viable future. As I noted in chapter 2, Germany's BDI industrial forum has estimated the costs of a German transition as 80% or 95% greenhouse gas reduction. It put them, respectively, at around €1.5 trillion and €2.3 trillion by 2050, or about 1.2% or 1.8% of Germany's GDP annually through to 2050, although with energy and cost savings from the programme reducing that substantially to around €470 billion or €960 billion respectively by 2050 (BDI 2018).

Consultants Wood MacKenzie have put the cost of reaching (and balancing) 100% renewable power supply in the United States at up to $4.5 trillion, if it had to be done by 2040, but less if done more slowly (ESJ 2019). That too is a large sum, but, as with the German figures, there would be savings. A University of San Francisco study claimed that the cost of transitioning the United States to 80% clean energy by 2050 might be equivalent to around 1% of GDP, but that this could be offset by the savings made by not having to buy fossil fuels (McMahon 2019).

Looking globally, leading financial consultants Morgan Stanley says a renewables programme aiming to supply 80% of the world's power by 2050 will require $14 trillion of investment and similar sums for other parts of the new energy system, including $20 trillion for establishing a hydrogen infrastructure (Klebnikov 2019).

These sums are huge and may be overstated but, as indicated above, they have to be set alongside the much larger costs from climate impacts that would be faced if no change were made. Moreover, quite apart from the problem of making assessments of costs so far ahead with new technologies and in the context of an unknown future environment, it is also important to point out that large sums would need to be spent anyway to replace existing power systems as they become obsolete.

It is also worth noting that there are different investment cost structures over time with renewables and conventional energy technology. With renewables, like wind and solar, you are investing in plants rather than fuel: in effect, you are buying a plant's lifetime's worth of energy up front. With regard to fossil plants, much of the cost is spread over their lifetime – in buying fuel – at unknown future cost. It may cost more initially to go for renewables but not when considered over the whole cost cycle or in terms of environmental costs. Energy-efficiency investment is similar in this respect to that for renewables but in some ways better. Once made, investment in, effectively, avoiding energy use ensures reduced energy costs and impacts into the future with, in most cases, no overheads or ongoing running costs.

So what is the bottom line on costs? I am not an economist but, from what I can see from the technical projections I have reviewed in this book, while there will be costs, on balance it does seem possible that a radical energy transition can be made without imposing impossibly high extra overall initial costs and it could lead to a system with lower running costs, all of this being offset by large savings. However, not everyone agrees and, as I have indicated, if we want to make the transition rapidly, it might initially cost more.

There are occasional episodes of panic when claims are made that very high costs will have to be faced. For example, the International Monetary Fund recently attracted media headlines by suggesting that a carbon tax set at the level it deemed necessary to contain climate change would lead to household energy bills rising 43% or more in some countries by 2030 (L. Elliott 2019).

The cumulative cost of not responding effectively to climate change may of course be very much higher, and the regulatory/state intervention approaches being promoted in the various New Green Deals *may* be less costly, disruptive and regressive than IMF's proposed high level of carbon tax. However, that is debatable. Bernie Sanders's ambitious climate plan for the United States, aiming for 100% of US power and transport needs with renewables by 2030 and then moving on to full decarbonization by 2050, has around $3 trillion spent on renewables/storage/grids, although it also includes other spending, for example about as much again on decarbonizing transport and on compensation for displaced workers, taking it up to $16.3 trillion in all.

That is a vast sum, but he claims that if no action on climate is taken the United States will lose $34.5 trillion in economic activity by the end of the century, whereas 'by taking bold and decisive action, we will save $2.9 trillion over ten years, $21 trillion over 30 years, and $70.4 trillion over 80 years' (Sanders 2019).

Arguments about the numbers and the politics will no doubt continue with, for example, provocatively, the total cost of a technology-led global climate-protection programme in the aforementioned Morgan Stanley study put at $50 trillion (Klebnikov 2019). Estimates like this may be excessive but there is an undoubted shock value in such pronouncements, even if inflated and ignoring potential savings. They highlight some of the key issues ahead.

If costs like this are deemed to be too high and/or Green New Deal programmes too politically scary, there are other, more extreme options for reducing emissions as noted in chapter 5, ranging from draconian social change to an end to growth. Would those who are unable or unwilling to face the cost of a green technology-led programme prefer that? Or are they clinging on to a belief that some lower-cost new technological breakthrough (or revived nuclear and maybe CCS) will come to the rescue? Or that climate change will go away?

The way ahead

As I have indicated, the energy transition will not be easy: it will present issues in some sectors and some countries, and it might have unwelcome short-term costs. A pessimistic view of the prospects for the wide adoption of renewables and energy-system change is that, given the uncertainties, change on the scale needed is impossible or at least can only be very slow. That view seems common in the fossil fuel industry, but nuclear supporters also sometimes share it: renewables cannot deliver fast enough. The alternative message seems to be that change is possible and can be achieved quite rapidly since, if money is not diverted into expensive nuclear projects, renewable expansion is able to ramp up fast. The new programme may have social and economic costs, but so will staying with the status quo, and any initial costs of change are well worth facing.

As far as I can see, that is right, and we have little choice. We have to stop burning fossil fuel as fast as possible. It will however take time, and even if we can get near to '100% renewables' by 2050, we will still be using fossil fuels for some while. In this book, I have not dealt much with how to handle that issue, although I have contributed a chapter to a recent book on managing the decline of fossil fuels in which I look critically at carbon capture (Elliott 2019d).

As I argued there, carbon capture and storage may have to be tried at some scale as an *interim* measure, but there are risks with focusing too much on this type of carbon abatement – it is not a long-term solution. We will not be able to continue to sweep our carbon emissions under the carpet indefinitely, finding room for ever-increasing amounts, if fossil fuel use is still expanding. At best, CCS can just buy time. At worst, investing in it can deflect and delay the development of renewables and upgraded efficiency, arguably the only long-term energy options for avoiding or limiting further climate change. In that context, as I argued in chapter 4, lumping carbon removal and carbon offsets in with zero-carbon renewables in 'net zero carbon' policies is fraught with problems: there are strategic conflicts (McLaren 2019).

As I noted in Box 4.5, progress on CCS has in fact been very slow, its costs being high and, although that may change, the prospects do not look good, with CCU perhaps being more likely to move ahead. However, that could lead to the production of more fuel to be burnt, creating CO_2. So it would not be overall carbon neutral. In addition, as I argued in chapter 4, power plants with added CCS or CCU cannot help much with grid balancing, so that the potential for enhancing their economics by offering this service may not be open to them.

Nevertheless, some analysts see the wide-scale development of negative emissions technologies (NETs) as being an emergency option. That does assume that CCS can be developed at scale. Moreover, quite apart from the land-use implications of BECCS and the energy needed for DACCS, there will also be CO_2 storage space/volume limits to how much NETs could help in trying to reduce CO_2 levels in the air. Arguably, we have already emitted too much. Certainly, there will not be enough space to store the CO_2 reliably and indefinitely if we continue to burn fossil fuel.

On balance, then, the potential for CCS/NETs may be limited. Some therefore look, in extremis, to geo-engineering as the ultimate emergency measure, including ways to block out sunlight. As I argued in chapter 5, it would surely make more sense to speed up the expansion of renewables and avoid risky and untested options like this.

I have also argued that the same applies to nuclear power: there are better options. In fact, it is perhaps odd that a technology that in 2017 only supplied around 2.2% of global final energy use (REN21 2019) has often managed to dominate the energy debate and indeed has even figured regularly in this book about renewables. Some do still look to the expansion of nuclear, but given its high cost and many other problems, including operational inflexibility and carbon emissions from its fuel production, I do not think nuclear will or should have much of a role and, as I have noted, I have contributed to or written books and studies on the case for a nuclear phase-out and associated issues (Elliott 2010, 2017b).

Instead of going over all that ground again, in this book I have focused on the more positive side and the case for ramping up renewables. However, implicit in this approach is that fossil fuels as well as nuclear have to step aside or, rather, be pushed aside: renewables should not be seen just as an additional option. In the book on managing fossil fuel decline mentioned above, one editor wrote:

> Simply put, there needs to be a decoupling of fossil fuels and renewable/low carbon energy; the latter cannot just simply pick up the slack of increasing demand or serve difficult to reach places. Unaddressed and unresolved, all this will do is serve to bloat the energy system with the same problems for issues such as climate change and energy security. (Wood 2019)

That is clearly true. Renewables must not become just a way to meet demand alongside fossil fuel (and that arguably also extends to nuclear), and the solution is also clear: demand has to be tamed and renewables have to take over supply. Hopefully, this book will have indicated how at least parts of that might be possible to achieve, rapidly, equitably and without undue cost or disruption via rapid decarbonization and via electrification of supply for most end uses, but also looking at non-electrical routes forward.

It also means improving end-use energy efficiency, something I have not followed up much in this book, in part since it is very sector and end-use specific. Nevertheless, despite the problems (see chapter 2), the potential for demand reduction via end-use technical measures is large, and their adoption needs to be accelerated (Eyre and Killip 2019).

While it is clear that we need renewables *and* energy efficiency, the relationship between adaptation and mitigation is more complicated. Some adaptation will be vital, as climate impacts hit, to reduce their local costs but, as I argued in chapter 6, there are strategic conflicts between adaptation and mitigation. Adaptation may be urgent, but the wider climate problem cannot be solved that way. It will just get worse unless emissions are cut (Schumacher 2019a).

As I argued above, it is the same with *carbon offsetting* via carbon removal and carbon sequestration systems. Some projects may nevertheless go ahead and, while there will be operational and CO_2 storage limits so that they cannot be relied on to deal with ever-expanding emissions into the far future, they could perhaps play an *interim* role, helping while renewables expand. Specifically, they could be used to offset the emissions resulting from the changeover process. In chapter 4, I argued that some interim use of fossil CCS and NETs might be made to compensate for the carbon debt of the fossil fuel and materials used during the initial renewable construction phase. For example, BECCS using bio-wastes, rather than specially grown energy crops, might be an option, avoiding new land use. DACCS would also avoid that problem.

However, I also suggested that planting more trees and adopting new farming practices were arguably better options for this interim compensatory role and indeed for carbon sequestration generally. There are certainly many additional environmental benefits associated with planting (and protecting) trees. Even so, there may be limits to the long-term sequestration potential of reforestation. Sadly, given the volume of CO_2 that would have to be stored, we would have to plant vast areas. Even if we stopped producing any more CO_2, given the spatial limitations, it is unclear whether natural carbon sequestration by trees and plants will be any more viable as carbon removal options in the *long term* than artificial sequestration, using negative emissions technologies like BECCS and DACCS: they all come up against spatial limits of one type or another (Elliott 2019d).

All of this arguably means that, amongst other things, personal or corporate carbon offset schemes, which for a fee offer to plant trees or to support other carbon removal options so as to compensate for carbon emissions (for example, from flying), may be increasingly problematic. As with CCS and NETs generally, they may help to buy time, limiting CO_2 impacts in the short term, but they do not deal with the CO_2 problem long term and at source. That can only be done by investing in zero-carbon generation or energy-saving projects (which, admittedly, some of the better offset schemes do) or, more directly, by not using energy, for example by not flying or driving.

Social and technical change

The latter choice highlights the general point that, while there are technical options for dealing with climate challenges, there are also social choice options, including using less energy. I have not spent much time exploring these social options in this book, but I did look at some of the issues, and how change might be brought about, in chapter 5. Some of the changes may be relatively easy and painless and may even improve life. Decentralization of some energy provisions may enhance local economies and communities but may also require some changes in lifestyle. So will some of the domestic energy-saving initiatives although possibly not as much as the changes needed in transport or in dietary habits, including less meat eating.

Attempts to reduce energy demand and carbon emissions would be aided if we could adopt a more sustainable approach to consumption. Some people already have voluntarily. However, more is needed. We must start decoupling consumption from prosperity as far as possible and also expand access to birth control measures to reduce population growth. We need to head for a sustainable global population level, another big, controversial topic.

For good or ill (it is well outside my area of expertise), I have not explored these large social issues in any detail nor the radical zero-growth policies some think may be necessary, but instead I have focused on what some call technical fixes, mainly on the energy supply side. Nevertheless, it has to be accepted that not all these technical supply-side solutions may be viable. It may also be impossible to tame energy demand, even with clever technical fixes and moderate social and behavioural changes. In that case, as I argued in chapter

5, more radical changes may be needed including, possibly, a move to lower economic growth.

As I have indicated, the more optimistic scenarios suggest that this will not be necessary. Renewables could, in theory, be expanded sufficiently to avoid the need for major social changes and to allow for continued economic growth. Not everyone will want that, and there are certainly good reasons, quite apart from climate issues, for moving away from relentless consumerism. However, for those living at or below the subsistence level, growth for now may be the only hope for a better future. Quite rightly, as I noted in chapter 6, they may ask why they are being denied the right to do what others have already done.

Nevertheless, we can still ask *what sort of growth* is needed: it does not necessarily have to be the same as at present or use the same technology, and it does not necessarily have to continue indefinitely. There are scenarios in which growth is slowed in developed countries but is condoned for a while, in 'catch-up' mode, in less developed countries, with aid provided by the rich countries to help them make the transition. Hopefully, that can enable the developing countries to 'leapfrog' the 'dirty technology' phase that the old industrial countries went through, with the transition in the developing countries increasingly being based on environmentally sound options.

Reaching global political agreements on plans like that, and even more so on population policy, may be beyond us, or at least may take some time, and some say we do not have the time. However, we might be able to make some movement in that direction, and, as I hope I have indicated, given the will to accept some changes, it should be possible to avoid extremes, with the rapid adoption of renewables and efficient end use providing part of the way forward.

There is a huge effort required to bring about the changes needed, which will require cooperation and collaboration across many separate areas of expertise and by countries across the world. However, while it can be hard to recognize, given our understandable worries about climate change and the sometimes dismal responses from some political leaders, in some ways, from the technological perspective, the prospects for the future do look quite good. Indeed, DNV

GL's energy group CEO has claimed that 'The technology is there if we want to make it happen. There is actually a toolbox available that you can work with if you really want to get there.' The real problem, he says, is that the policy support is lacking (Merchant 2019).

It may be wise not to oversell technological solutions (we need more than just technical fixes) and it is easy to be dismissive of specific renewable energy projects and systems, taken in isolation, setting up 'straw men' for one-liner ridicule, and there is no shortage of such efforts (Mills 2019). However, although it can be overstated, the reality seems to be that, while the climate threat is significant and worrying, technology moves on, better systems emerge and the prospects for a successful energy transition continually improve.

The CEO of the UK Climate Change Committee put it like this:

a transition to a near zero carbon economy is now technically achievable – credible scenarios now exist to achieve near-full decarbonisation in most sectors. This is genuine progress. Electrification with zero carbon supply takes us much of the way – and there are now credible alternatives, like hydrogen, for those applications where that strategy won't work. And even in those sectors where emissions look set to continue, we can match emissions with greenhouse gas removals. So it is possible. But that does not mean it is feasible. The scale of the change is enormous, and this transition must take place at remarkable (although not unprecedented) speed. (Stark 2019a)

That was from a UK perspective, and the United Kingdom is blessed with some excellent renewable resources (so CCS/carbon removal should actually not be needed), but the picture is similar elsewhere, including in countries with much lower population densities – and more sunshine! Political changes will need to happen to make it a reality across the world, and social change will also be needed, some of it radical. However, there should be social and political pay-offs, as well as environmental benefits, beyond just limiting climate change.

Launching IRENA's publication 'A New World: The Geopolitics of the Energy Transformation', IRENA's director general said that the transition to renewables and away from fossil fuels is 'a move away from the politics of scarcity and conflict to abundance and peace with new opportunities for many countries' (IRENA 2019c). And in broad terms that is hard to gainsay.

To my mind, it harks back to what Adlai Stevenson said in a famous speech to the UN in 1965: 'We travel together, passengers on a little spaceship, dependent on its vulnerable reserves of air and soil; all committed for our safety to its security and peace; preserved from annihilation only by the care, the work, and, I will say, the love we give our fragile craft' (Stevenson 1965).

Tragically, the global ecological situation has in many ways worsened since 1965 for ourselves and for the other passengers on Adlai Stevenson's 'little spaceship' – all the other species, plants and life forms on whom our future, and the future of our descendants, depend. However, to some extent, the current climate and environmental crisis makes what has to be done a little clearer, at least in terms of energy, with the benefits of change also being clearer. Although, as I hope I have indicated, it will not be an easy transition, there would seem to be no realistic or acceptable alternative. And we have to get on with it as fast as possible.

Afterword: on updates and wider approaches

There is a lot going on in the renewable energy development and sustainable energy policy fields at present. If you need regular updates, you may find my weekly 1,000-word blog posts and free bimonthly e-newsletter *Renew On Line* helpful. Both are accessible at https://renewnatta.wordpress. com.

My focus in this book is mainly on technology, as one response to climate and environmental problems. As I have indicated, it is only one of several possible approaches. I could have engaged more with the wide-ranging debate on climate policy, social and economic development and so on. I do touch on these wider issues, and clearly any resolution of our climate and environmental problems will require a multifaceted approach, drawing on expertise from a range of disciplinary traditions and crossing boundaries. However, I contend that a technological focus is an important element: too often policy decisions and public debates are based on a poor understanding of what is technologically possible (or impossible). Equally, some technologists are prone to dismissing policy issues as, in effect, irritants, getting in the way of technical progress. More generally, engineers, climate scientists, economists and political scientists and so on can all end up in separate, specialist, jargon-entrenched silos. We all need to talk more with activists as well as with everyone else.

Some possible starting points: http://www.zedbooks. net/shop/book/climate-futures/; https://www.sciencedirect. com/journal/energy-research-and-social-science/; and even this: http://www.resilience.org/stories/2019-09-12/what-is-energy-denial. And finally, in contrarian mood, a fine attempt at a refutation of just about all green energy arguments: https://oilprice.com/Alternative-Energy/Renewable-Energy/ Renewable-Energys-Inconvenient-Truth.html.

References

Web link references: all web URLs were accessed and live in October 2019.

Abdulla, A. (2018) 'The demise of US nuclear power in 4 charts'. *The Conversation*, 1 August. https://theconversation.com/the-demise-of-us-nuclear-power-in-4-charts-98817

Aggidis, G. (2017) 'Spinning sail technology is poised to bring back wind-powered ships'. *The Conversation*, 21 March. https://theconversation.com/spinning-sail-technology-is-poised-to-bring-back-wind-powered-ships-74872

Aghahosseini, A., Bogdonov, D., Barbosa, L. and Breyer, C. (2019) 'Analysing the feasibility of powering the Americas with renewable energy and inter-regional grid interconnections by 2030'. *Renewable and Sustainable Energy Reviews* 105: 187–205. https://www.sciencedirect.com/science/article/pii/S1364032119300504

Ahl, A., Yarime, M., Tanaka, K. and Sagawa, D. (2019) 'Review of blockchain-based distributed energy: Implications for institutional development'. *Renewable and Sustainable Energy Reviews* 107: 200–11. http://www.sciencedirect.com/science/article/pii/S1364032119301352

Amadeo, K. (2019) 'The Green New Deal and why it's happening now'. *The Balance*, 1 August. https://www.thebalance.com/green-new-deal-4582071

Ambrose, J. (2019) 'Cool running: supermarket fridges could help power UK'. *The Guardian*, 23 June. https://www.theguardian.com/business/2019/jun/23/cool-running-supermarket-fridges-could-help-power-uk

Anderson, N. (2019) 'Can the energy industry rise to the challenge of climate change?'. Wood Mackenzie. http://

www.woodmac.com/news/feature/can-the-energy-industry-rise-to-the-challenge-of-climate-change/

Aris, C. (2018) 'A cheaper, cleaner electricity system'. London: Global Warming Policy Foundation. http://www.thegwpf.org/content/uploads/2019/01/Capell-Aris-UK-Electricity-System.pdf

Arrobas, D., Hund, K., McCormick, M., Ningthoujam, J. and Drexhage, J. (2017) 'The growing role of minerals and metals for a low carbon future'. The World Bank. http://documents.worldbank.org/curated/en/207371500386458722/The-Growing-Role-of-Minerals-and-Metals-for-a-Low-Carbon-Future

Atkin, E. (2019) 'Some of the biggest green groups have cold feet over the "Green New Deal"'. *New Republic*, 15 January. https://newrepublic.com/article/152885/biggest-green-groups-cold-feet-green-new-deal

Auffhammer, M. (2019) 'Vastly more travel'. Energy Institute at Haas blog, University of Berkeley, CA, 4 March. https://energyathaas.wordpress.com/2019/03/04/_v_astly-_m_ore-_t_ravel/

Aurora (2018) 'Power sector modelling: System cost impact of renewables'. Report for the National Infrastructure Commission, Aurora Energy Research: https://www.nic.org.uk/wp-content/uploads/Power-sector-modelling-final-report-1-Aurora-Energy-Research.pdf

Baculinao, E. (2016) 'China unveils proposal for $50 trillion global electricity network'. NBC News, 31 March. https://www.nbcnews.com/business/energy/china-unveils-proposal-50-trillion-global-electricity-network-n548376

Bairstow, J. (2019a) 'Smarter and more flexible energy system "could save UK up to £40bn"'. *Energy Live News*, 17 July. https://www.energylivenews.com/2019/07/17/smarter-and-more-flexible-energy-system-could-save-uk-up-to-40bn/

Bairstow, J. (2019b) 'Biogas "could slash global greenhouse gas emissions by up to 13%"'. *Energy Live News*, 4 July. https://www.energylivenews.com/2019/07/04/biogas-could-slash-global-greenhouse-gas-emissions-by-up-to-13/

Bairstow, J. (2019c) 'Energy workers "must be supported through renewable transition"'. *Energy Live News*, 11 March. https://www.energylivenews.com/2019/03/11/energy-workers-must-be-supported-through-renewable-transition/

Baños Ruiz, I. (2018) 'Hydropower supply dries up with climate change'. Deutsche Welle, 1 March. https://www.dw.com/en/hydropower-supply-dries-up-with-climate-change/a-42472070

Baraniuk, C. (2017) 'Future energy: China leads world in solar power production'. BBC News, 22 June. https://www.bbc.co.uk/news/business-40341833

Barron, A. (2018) 'Why countries with biggest renewable reserves will become superpowers of tomorrow'. *The Independent*, 22 February. https://www.independent.co.uk/environment/renewable-superpowers-fossil-fuel-era-over-reserves-lithium-copper-rare-metals-solar-energy-a8217786.html

Bastani, A. (2019) *Fully Automated Luxury Communism*. London and New York: Verso.

Bazilian, M., Bradshaw, M., Goldthau, A. and Westphal, K. (2019) 'Model and manage the changing geopolitics of energy'. *Nature*, 1 May. https://www.nature.com/articles/d41586-019-01312-5

BBC (2018) 'Climate change: Is your Netflix habit bad for the environment?'. BBC Reality Check team, 12 October. https://www.bbc.co.uk/news/technology-45798523

BDI (2018) 'Climate paths for Germany'. BDI Industry Forum. https://english.bdi.eu/media/presse/presse/downloads/20180308_Climate_Paths_for_Germany_ExecutiveSummary_FINAL.pdf

Beam (2019) '100% renewable energy across Europe is more cost-effective than the current energy system'. *The Beam*, 8 February. https://cleantechnica.com/2019/02/08/proven-100-renewable-energy-across-europe-is-more-cost-effective-than-the-current-energy-system/

Becker, S., Demski, C., Evensen, D. and Pidgeon, N. (2019) 'Of profits, transparency, and responsibility: Public views on financing energy system change in Great Britain'. *Energy Research & Social Science* 55: 236–46. https://www.sciencedirect.com/science/article/pii/S2214629618311988

Bedeschi, B. (2019) 'Offshore wind hydrogen could be subsidy-free within 10 years'. New Energy Update, 1 May. http://newenergyupdate.com/wind-energy-update/offshore-wind-hydrogen-could-be-subsidy-free-within-10-years

BEIS (2018a) 'Hydrogen for heating: Atmospheric impacts'. London: Department for Business, Energy and Industrial Strategy. https://assets.publishing.service.gov.uk/government/uploads/system/uploads/attachment_data/file/760538/Hydrogen_atmospheric_impact_report.pdf

BEIS (2018b) 'Digest of United Kingdom energy statistics 2018' (DUKES) 2018: Main report'. London: Department for Business, Energy and Industrial Strategy. https://www.gov.uk/government/statistics/digest-of-uk-energy-statistics-dukes-2018-main-report

BEIS (2019a) 'Energy public attitude tracker'. Wave 29. London: Department for Business, Energy and Industrial Strategy. https://assets.publishing.service.gov.uk/government/uploads/system/uploads/attachment_data/file/800429/BEIS_Public_Attitudes_Tracker_-_Wave_29_-_key_findings.pdf

BEIS (2019c) 'Proposals to protect consumers whilst guaranteeing payments for households with solar by unlocking smarter energy system'. London: Department for Business, Energy and Industrial Strategy. http://www.gov.uk/government/news/proposals-to-protect-consumers-whilst-guaranteeing-payments-for-households-with-solar-by-unlocking-smarter-energy-system

Beisner, C. (2019) 'Does fighting global warming help or hurt the poor?'. *Patriot Post*, 26 January. https://patriotpost.us/opinion/60789-does-fighting-global-warming-help-or-hurt-the-poor?

Berkeley Lab (2018) 'Utility-scale solar'. Berkeley Lab report. https://emp.lbl.gov/utility-scale-solar/

Berkeley Lab (2019) 'Tracking the sun'. Berkeley Lab report. https://emp.lbl.gov/tracking-the-sun/

Bischof-Niemz, T. and Creamer, T (2018) *South Africa's Energy Transition: A Roadmap to a Decarbonised, Low-Cost and Job-Rich Future*. London: Routledge.

Blakers, A., Stocks, M., Lu, B., Cheng, C., and Nadolny, A. (2019) 'Global pumped hydro atlas'. Australian National University. http://re100.eng.anu.edu.au/global/

BNEF (2018a) 'New energy outlook 2018'. Bloomberg New Energy Finance. https://about.bnef.com/new-energy-outlook/#toc-download

BNEF (2018b) 'Runaway 53GW solar boom in China pushed global clean energy investment ahead in 2017'. Bloomberg New Energy Finance. https://about.bnef.com/blog/runaway-53gw-solar-boom-in-china-pushed-global-clean-energy-investment-ahead-in-2017/

BNEF (2019) 'Clean energy investment exceeded $300 billion once again in 2018'. Bloomberg New Energy Finance, 16 January. https://about.bnef.com/blog/clean-energy-investment-exceeded-300-billion-2018/

Bogdanov, D. and Breyer, C. (2016) 'North-East Asian Super Grid for 100% renewable energy supply: Optimal mix of energy technologies for electricity, gas and heat supply options', *Energy Conversion and Management* 112: 176–90. http://www.sciencedirect.com/science/article/pii/S0196890416000364

Bogdanov, D., Child, M. and Breyer, C. (2019) 'Reply to "Bias in energy system models with uniform cost of capital assumption"'. *Nature Communications*. https://www.nature.com/articles/s41467-019-12469-y.pdf

Bogdanov, D., Farfan, J., Sadovskaia, K. et al. (2019) 'Radical transformation pathway towards sustainable electricity via evolutionary steps', *Nature Communications* 10, Article number: 1077. https://www.nature.com/articles/s41467-019-08855-1

Boucher, J. L. and Heinonen, J. (eds) (2019) *Sustainable Consumption, Promise or Myth? Case Studies from the Field.* Newcastle upon Tyne: Cambridge Scholars publishing. http://www.cambridgescholars.com/sustainable-consumption-promise-or-myth-case-studies-from-the-field

Bouckaert, S. and Goodson, T. (2019) 'Commentary: The mysterious case of disappearing electricity demand'. Paris: International Energy Agency. https://www.iea.org/newsroom/news/2019/february/the-mysterious-case-of-disappearing-electricity-demand.html

Bowler, T. (2019) 'Why the age of electric flight is finally upon us'. BBC News, 3 July. https://www.bbc.co.uk/news/business-48630656

Bown, J. (2019) 'Could high-flying drones power your home one day?'. BBC Business News, 3 May. https://www.bbc.co.uk/news/business-48132021

Boysen, L., Lucht, W., Gerten, D., Heck, V., Lenton, T. and Shellnhuber, H. J. (2017) 'The limits to global-warming mitigation by terrestrial carbon removal'. *Earth's Future* 5, 463–74. https://agupubs.onlinelibrary.wiley.com/doi/epdf/10.1002/2016EF000469

BP (2018a) 'BP statistical review of world energy'. BP. https://www.bp.com/content/dam/bp/business-sites/en/global/corporate/pdfs/energy-economics/statistical-review/bp-stats-review-2018-full-report.pdf

BP (2018b) 'Energy outlook'. BP. https://www.bp.com/en/global/corporate/energy-economics/energy-outlook/energy-outlook-downloads.html

BP (2019) 'Energy outlook'. BP. https://www.bp.com/content/dam/bp/business-sites/en/global/corporate/pdfs/energy-economics/energy-outlook/bp-energy-outlook-2019.pdf

Brack, D. (2017) 'Woody biomass for power and heat: Impacts on the global climate'. London: Chatham House. https://www.chathamhouse.org/publication/woody-biomass-power-and-heat-impacts-global-climate

Bradford, J. (2019) *The Future is Rural.* Corvallis, OR: Post Carbon Institute. http://www.postcarbon.org/publications/the-future-is-rural/

Brand, C., Anable, J. and Morton, C. (2019) 'Energy for mobility: Exploring systemic change in a "net zero" world'. UK Energy Research Centre report, 26 June. http://www.ukerc.ac.uk/publications/team-energy-for-mobility.html

Brattle Group (2015) 'Comparative generation costs of utility-scale and residential-scale PV in Xcel Energy Colorado's service

area'. Brattle Group. https://brattlefiles.blob.core.windows.net/
system/publications/pdfs/000/005/188/original/comparative_
generation_costs_of_utility-scale_and_residential-scale_pv_in_
xcel_energy_colorado's_service_area.pdf?1436797265

Breyer, M. (2019) 'Biggest dam removal in European history has
begun with the Vezins dam'. *Tree Hugger*, 17 June. http://www.
treehugger.com/environmental-policy/biggest-dam-removal-
european-history-begins-vezins-dam.html

Brook, B. et al. (2015) 'An ecomodernist manifesto' (produced by
18 academics). http://www.ecomodernism.org/manifesto/

Brown, D. (2018) 'Carbon-constrained scenario 2018'. Wood
Mackenzie. https://www.woodmac.com/reports/macroeconomics-
risks-and-global-trends-carbon-constrained-scenario-2018-
navigating-a-challenging-path-to-lower-global-emissions-32846/

Brown, G. (2018) '5 reasons blockchain is game-changing for
solar energy'. Aurora blog, 21 February. https://blog.aurorasolar.
com/5-reasons-blockchain-is-game-changing-for-solar-energy

Brummer, V. (2018) 'Community energy – benefits and barriers'.
Renewable and Sustainable Energy Reviews 94: 187–96: https://
www.sciencedirect.com/science/article/pii/S1364032118304507

Burke, J. and Byrnes, R. (2019) 'What the UK can learn from
carbon pricing schemes around the world'. *Carbon Brief*, 2
August. https://www.carbonbrief.org/guest-post-what-the-uk-
can-learn-from-carbon-pricing-schemes-around-the-world

Burkhardt, P. and Vecchiatto, P. (2018) 'South Africa drops
nuclear, adds renewables in energy plan'. Bloomberg, 27
August. https://www.bloomberg.com/news/articles/2018-08-27/
south-africa-drops-nuclear-power-in-new-energy-capacity-plans

Bush, G. W. (2001) 'Text of a letter from the President to Senators
Hagel, Helms, Craig, and Roberts'. 13 March. https://georgewbush-
whitehouse.archives.gov/news/releases/2001/03/20010314.html

Calzadilla, A., Wiebelt, M., Blohmke, J. and Klepper, G. (2014)
'Desert Power 2050: Regional and sectoral impacts of renewable
electricity production in Europe, the Middle East and North
Africa'. Kiel Working Paper No. 1891. http://www.gci.org.uk/
Documents/KWP+1891.pdf

Caradonna, J., Heinberg, R., Borowy, I., et al. (2019) 'A call to
look past *An Ecomodernist Manifesto*: A degrowth critique'.
Resilience statement with 18 authors and endorsers. https://www.
resilience.org/wp-content/uploads/articles/General/2015/05_
May/A-Degrowth-Response-to-An-Ecomodernist-Manifesto.pdf

Carbon Brief (2018) 'The impacts of climate change at 1.5C, 2C
and beyond', *Carbon Brief*. https://interactive.carbonbrief.org/
impacts-climate-change-one-point-five-degrees-two-degrees/

Carbon Brief (2019) 'Mapped: The world's coal power plants'. *Carbon Brief Infographics*, 25 March. https://www.carbonbrief. org/mapped-worlds-coal-power-plants

Carbon Tracker (2019) 'The political tipping point: Why the politics of energy will follow the economics'. *Carbon Tracker*, 22 January. https://www.carbontracker.org/the-political-tipping-point/

Carrington, D. (2018) 'Solar geoengineering could be "remarkably inexpensive" – report'. *The Guardian*, 23 November. https://www.theguardian.com/environment/2018/nov/23/solar-geoengineering-could-be-remarkably-inexpensive-report

Casey, T. (2017) 'The Desertec Sahara solar dream didn't die after all – It's baaaack . . .'. *Clean Technica*, 11 August. https://cleantechnica.com/2017/08/11/desertec-sahara-solar-dream-didnt-die-baaaack/

CAT (2013) 'Zero carbon: Rethinking the future'. Machynlleth: Centre for Alternative Technology. https://www.cat.org.uk/info-resources/zero-carbon-britain/research-reports/zero-carbon-rethinking-the-future/

CCC (2018a) 'Biomass in a low-carbon economy'. London: Committee on Climate Change. http://www.theccc.org.uk/wp-content/uploads/2018/11/Biomass-in-a-low-carbon-economy-CCC-2018.pdf

CCC (2018b) 'Hydrogen in a low-carbon economy'. London: Committee on Climate Change. https://www.theccc.org.uk/wp-content/uploads/2018/11/Hydrogen-in-a-low-carbon-economy.pdf

CCC (2019) 'Net zero – the UK's contribution to stopping global warming'. London: Committee on Climate Change. http://www.theccc.org.uk/publication/net-zero-the-uks-contribution-to-stopping-global-warming/

CEMAC (2017) Benchmark analysis. US Clean Energy Manufacturing Analysis Centre. https://www.manufacturingcleanenergy.org/images/cemac-benchmarks-figures/es-3.jpg

Chandran, R. (2019) 'Why floating solar panels are on the rise'. World Economic Forum blog, 15 February. https://www.weforum.org/agenda/2019/02/in-land-scarce-southeast-asia-solar-panels-float-on-water

Chestney, N. (2018) 'Bioenergy leads growth in renewable energy consumption to 2023: IEA'. *Reuters*, 8 October. https://uk.reuters.com/article/us-iea-renewables/bioenergy-leads-growth-in-renewable-energy-consumption-to-2023-iea-idUKKCN1MH123

Clack, C., Qvist, S., Apt, J., Bazilian, M., Brandt, A., Caldeira, K., Davis, S., Diakov, V., Handschy, M., Hines, P., Jaramillo, P., Kammen, D., Long, J., Morgan, G., Reed, A., Sivaram, V., Sweeney, J., Tynan, G., Victor, D., Weyant, J. and Whitacre, J.

(2017) 'Evaluation of a proposal for reliable low-cost grid power with 100% wind, water, and solar'. *PNAS* 114(26): 6722–7. http://www.pnas.org/content/114/26/6722

Clark, B. (2018) 'Renewable energy will be "effectively free" by 2030'. The Next Web, 14 August. https://thenextweb.com/insider/2018/08/14/analyst-renewable-will-be-effectively-free-by-2030/

Climate Change letter (2019) Text of letter to Congress from 625 green groups. https://www.biologicaldiversity.org/programs/climate_law_institute/legislating_for_a_new_climate/pdfs/Letter-to-Congress-%20Legislation-to-Address-the-Urgent-Threat-of-Climate-Change.pdf

Colagrossi, M. (2019) 'The dirty side of renewable energy'. *Big Think*, 29 July. https://bigthink.com/technology-innovation/renewable-energy-dirty-mining

Courvoisier, A. (2019) 'Techs for the future I: The false promises of "green growth"'. Reflexions on Ethics, 11 March blog. http://ethicsinstem.blogspot.com/2019/03/techs-for-future-i-false-promises-of.html

CWIF (2019) 'Summary of wind turbine accident data to 31st March 2019'. Caithness Windfarm Information Forum. http://www.caithnesswindfarms.co.uk/AccidentStatistics.htm

Dale, G. (2019) 'Climate, communism and the Age of Affluence?'. *The Ecologist*, 29 May. https://theecologist.org/2019/may/29/climate-communism-and-age-affluence

Davis, L. (2017a) 'Are Mexican renewables really this cheap?'. Berkeley, CA: University of California Energy Institute at Haas blog, 4 December. https://energyathaas.wordpress.com/2017/12/04/are-mexican-renewables-really-this-cheap/

Davis, L. (2017b) 'Evidence of a decline in electricity use by US Households'. Berkeley, CA: University of California Energy Institute at Haas, 8 May. https://energyathaas.wordpress.com/2017/05/08/evidence-of-a-decline-in-electricity-use-by-u-s-households/

DECC (2014) 'Life cycle impacts of biomass electricity in 2020'. London: Department of Energy and Climate Change. https://assets.publishing.service.gov.uk/government/uploads/system/uploads/attachment_data/file/349024/BEAC_Report_290814.pdf

Deign, J. (2014) 'Desertec: Slow death or healthy evolution?'. New Energy Update, 3 November. http://analysis.newenergyupdate.com/csp-today/markets/desertec-slow-death-or-healthy-evolution?

Deign, J. (2018) 'Gloomy prospects in IEA's latest world energy outlook'. Greentech Media, 13 November. https://www.

greentechmedia.com/articles/read/iea-latest-world-energy-outlook-fossil-fuels-renewables#gs.l8fm15

Delucchi, M. and Jacobson, M. (2012) 'Response to "A critique of Jacobson and Delucchi's proposals for a world renewable energy supply" by Ted Trainer'. *Energy Policy* 44: 482–4. http://www.sciencedirect.com/science/article/pii/S0301421511008731

Delucchi, M. and Jacobson, M. (2016) 'Meeting the world's energy needs entirely with wind, water, and solar power'. *Bulletin of the Atomic Scientists* 69(4): 30–40. http://bos.sagepub.com/content/69/4/30.full

Desertec (2011) Desertec-EU-MENA map. Wikipedia archive source. https://commons.wikimedia.org/wiki/File:DESERTEC-Map_large.jpg

Desertec (2019) Desertec Foundation. http://www.desertec.org

Diesendorf, M. and Elliston, B. (2018) 'The feasibility of 100% renewable electricity systems: A response to critics'. *Renewable and Sustainable Energy Reviews* 93: 318–30. https://www.sciencedirect.com/journal/renewable-and-sustainable-energy-reviews/vol/93/suppl/C

Dii (2012) 'Desert Power 2050'. Desertec Industrial Initiative. http://www.desertec-uk.org.uk/reports/DII/DPP_2050_Study.pdf

DOE (2018a) 'Wind technologies market report'. Berkeley Lab report for the US Department of Energy. https://emp.lbl.gov/wind-technologies-market-report

DOE (2018b) '2017 offshore wind technologies market update'. US Department of Energy. https://www.energy.gov/sites/prod/files/2018/09/f55/71709_V4.pdf

Dong, W. and Ye, Q. (2018) 'Utility of renewable energy in China's low-carbon transition'. Washington, DC: Brookings Institution, 18 May. https://www.brookings.edu/2018/05/18/utility-of-renewable-energy-in-chinas-low-carbon-transition/

Drake Landing Solar Community (2019) Drake Landing solar storage project. Canada. http://www.dlsc.ca

DUKES (2019) Digest of UK Energy Statistics, Department for Business, Energy and Industrial Strategy, London. https://www.gov.uk/government/statistics/digest-of-uk-energy-statistics-dukes-2019

DW (2018) 'Saudi Arabia puts world's biggest solar power project on hold'. Deutsche Welle, 1 October. http://www.dw.com/en/saudi-arabia-puts-worlds-biggest-solar-power-project-on-hold/a-45706685

Dyson, M. (2019) 'A bridge backward? The risky economics of new natural gas infrastructure in the United States'. Rocky Mountain Institute. https://rmi.org/a-bridge-backward-the-risky-economics-of-new-natural-gas-infrastructure-in-the-united-states

EACAS (2018) 'Negative emission technologies: What role in meeting Paris Agreement targets?'. Halle: European Academies' Science Advisory Council. https://easac.eu/publications/details/easac-net/

EASA (2019) European Aviation Safety Agency, Emissions: Overview of Aviation sector'. https://www.easa.europa.eu/eaer/topics/overview-aviation-sector/emissions

EC (2018) 'A clean planet for all: A European strategic long-term vision for a prosperous, modern, competitive and climate–neutral econmy'. European Commission. htts://eur-lex.europa.eu/legal-content/EN/TXT/?uri=CELEX:52018DC0773

EC (2019) 'Europeans' attitudes on EU energy policy and to climate change (reports 490 and 492). European Commission. https://ec.europa.eu/commfrontoffice/publicopinion/index.cfm/Survey/index#p=1&instruments=SPECIAL/surveyKy/2212

ECI (2018) 'ECI to lead new £19 million research centre on energy demand'. https://www.eci.ox.ac.uk/news/2018/0326.html

Egli, F., Steffen, B. and Schmidt, T. (2019) 'Bias in energy system models with uniform cost of capital assumption'. *Nature Communications.* https://www.nature.com/articles/s41467-019-12468-z.pdf

EIA (2019) 'Annual energy outlook 2019'. Washington, DC: Energy Information Administration. http://www.eia.gov/outlooks/aeo/

Element Energy and E4Tech (2018) 'Cost analysis of future heat infrastructure options'. Report for National Infrastructure Commission, Element Energy and E4Tech. http://www.nic.org.uk/wp-content/uploads/Element-Energy-and-E4techCost-analysis-of-future-heat-infrastructure-Final.pdf

Elliott, D (2004) 'Energy efficiency and renewables'. *Energy and Environment* 15(6): 1099–1105. https://journals.sagepub.com/doi/10.1260/0958305043026636

Elliott, D. (ed.) (2010) *Nuclear or Not?* Basingstoke: Palgrave Macmillan.

Elliott, D. (2013a) 'Emergence of European supergrids – Essay on strategy issues'. *Energy Strategy Reviews* 1(3): 171–3. https://www.sciencedirect.com/science/article/pii/S2211467X12000120

Elliott, D. (2013b) *Fukushima: Impacts and Implications.* Basingstoke: Palgrave Macmillan.

Elliott, D. (2015) 'Green jobs and the ethics of energy', in M. Hersh (ed.), *Ethical Engineering for International Development and Environmental Sustainability.* London: Springer.

Elliott, D. (2016) *Balancing Green Power.* Bristol: Institute of Physics Publishers.

Elliott, D. (2017a) 'Energy storage systems'. Institute of

Physics Publishers, free download. http://iopscience.iop.org/book/978-0-7503-1531-9

Elliott, D. (2017b) *Nuclear Power: Past, Present and Future*. San Rafael, CA: Morgan & Clay pool Publishers.

Elliott, D. (2019a) *Renewables: A Review of Sustainable Energy Supply Options*. Bristol: IOP Publishers.

Elliott, D. (2019b) *Renewables Energy in the UK: Past, Present and Future*. Basingstoke: Palgrave Macmillan.

Elliott, D. (2019c) 'Fusion – some new issues'. Renew Extra blog, 1 June. https://newrenewextra.blogspot.com/2019/06/fusion-some-new-issues.html

Elliott, D. (2019d) 'Carbon capture and renewables: Strategic conflicts or tactical complementarities', in G. Wood and K. Baker (eds), *The Palgrave Handbook of Managing Fossil Fuels and Energy Transitions*. Basingstoke: Palgrave Macmillan.

Elliott, D. (2019e) 'Cities and renewable energy', in S. Shmelev (ed.), *Sustainable Cities Reimagined*. London: Routledge.

Elliott, D. (forthcoming) 'Technology and the future', in D. Uzzell, D. Stevis and N. Räthzel (eds), *Handbook on Environmental Labour Studies*. Basingstoke: Palgrave Macmillan.

Elliott, D. and Cook, T. (2018) *Renewable Energy: From Europe to Africa*. Basingstoke: Palgrave Macmillan.

Elliott, D. and Cook, T. (2019) 'Renewable energy in Africa: Changing support systems', in A. Sayigh (ed.), *Renewable Energy and Sustainable Buildings*. WREC 2018 selected papers. Charn: Springer.

Elliott, L. (2019) 'Energy bills will have to rise sharply to avoid climate crisis, says IMF'. *The Guardian*, 10 October. https://www.theguardian.com/environment/2019/oct/10/carbon-taxes-have-to-rise-sharply-to-avoid-climate-crisis-says-imf

Ellsmoor, J. (2019a) 'Innovation takes to the skies: Electric planes are about to revolutionize the airline industry'. *Forbes*, 7 March. http://www.forbes.com/sites/jamesellsmoor/2019/03/07/innovation-takes-to-the-skies-electric-planes-are-about-to-revolutionize-the-airline-industry

Ellsmoor, J. (2019b) 'Under Trump's tariffs, the US lost 20,000 solar energy jobs'. *Forbes*, 24 February. http://www.forbes.com/sites/jamesellsmoor/2019/02/24/under-trumps-tariffs-the-us-lost-20000-solar-energy-jobs/

Enevoldsen, P., Permien, F-H., Bakhtaoui, I., et al. (2019) 'How much wind power potential does Europe have? Examining European wind power potential with an enhanced socio-technical atlas'. *Energy Policy* 132: 1092–1100. https://www.sciencedirect.com/science/article/pii/S0301421519304343

EPIC (2019) 'Do renewable portfolio standards deliver?'. Energy Policy Institute at the University of Chicago working paper. https://epic.uchicago.edu/research/publications/do-renewable-portfolio-standards-deliver

ESJ (2019) 'Decarbonizing US power grid to cost $4tr+, 900GW of energy storage needed'. *Energy Storage Journal*, 25 July. http://www.energystoragejournal.com/2019/07/25/decarbonizing-us-power-grid-to-cost-4tr-900gw-of-energy-storage-needed/

ETI (2016) 'Public perceptions of bioenergy in the UK'. Loughborough: Energy Technologies Institute. http://www.eti.co.uk/insights/public-perceptions-of-bioenergy-in-the-uk

ETI (2017) 'Carbon capture and storage needs to be part of the UK's future energy system or we will face "substantially higher costs"'. *Politics Home*, 4 April. https://www.politicshome.com/news/uk/energy/opinion/energy-technologies-institute/84830/carbon-capture-and-storage-needs-be-part/

Equinor (2019) 'Market outlook for floating offshore wind'. Equinor. https://www.equinor.com/en/what-we-do/hywind-where-the-wind-takes-us/the-market-outlook-for-floating-offshore-wind.html

ER (2019) Extinction Rebellion campaign web site. https://rebellion.earth/

EUCERS (2018) 'The future of energy and climate security'. Kings College, London: European Centre for Energy and Resource Security. https://www.europeangashub.com/wp-content/uploads/2018/09/strategy-paper-17.pdf

Evans, S. (2015) 'Investigation: Does the UK's biomass burning help solve climate change?'. *Carbon Brief*, 11 May. https://www.carbonbrief.org/investigation-does-the-uks-biomass-burning-help-solve-climate-change

Evans, S. (2017) 'Factcheck: How much power will UK electric vehicles need?'. *Carbon Brief*, 13 July. https://www.carbonbrief.org/factcheck-how-much-power-will-uk-electric-vehicles-need

Evans, S. (2019a) 'Analysis: UK electricity generation in 2018 falls to lowest level since 1994'. *Carbon Brief*, 3 January. http://www.carbonbrief.org/analysis-uk-electricity-generation-2018-falls-to-lowest-since-1994

Evans, S. (2019b) 'Analysis: Renewables could match coal power within 5 years, IEA reveals'. *Carbon Brief*, 21 October. https://www.carbonbrief.org/analysis-renewables-could-match-coal-power-within-5-years-iea-reveals

Exxon Mobil (2018) '2018 outlook for energy: A view to 2040'. Exxon Mobil. https://www.aop.es/wp.content/uploads/2019/05/2018-Outlook-for-Energy-Exxon.pdf

Eyre, N. and Killip, G. (eds) (2019) 'Shifting the focus: Energy demand in a net-zero carbon UK'. Oxford: Centre for Research into Energy Demand Solutions. http://www.creds.ac.uk/creds-launches-first-major-report-shifting-the-focus-energy-demand-in-a-net-zero-carbon-uk

Fairley, P. (2019) 'China's ambitious plan to build the world's biggest supergrid'. *IEEE Spectrum*, 21 February. https://spectrum.ieee.org/energy/the-smarter-grid/chinas-ambitious-plan-to-build-the-worlds-biggest-supergrid

Farand, C. (2019) 'Preparations for the climate crisis will save trillions, commission finds'. *Climate Home News*, 10 September. https://www.climatechangenews.com/2019/09/10/preparations-climate-crisis-will-save-trillions-commission-finds/

Farfan, J. and Breyer, C. (2018) 'Combining floating solar photovoltaic power plants and hydropower reservoirs: A virtual battery of great global potential'. *Energy Procedia* 155: 403–11. https://www.sciencedirect.com/science/article/pii/S1876610218309858

Fasihi, M., Efimova, O. and Breyer, C. (2019) 'Techno-economic assessment of CO_2 direct air capture plants'. *Journal of Cleaner Production* 224: 957–80. https://www.sciencedirect.com/science/article/pii/S0959652619307772?via%3Dihub#!

Firestone, J. and Kirk, H. (2019) 'A strong relative preference for wind turbines in the United States among those who live near them'. *Nature Energy* 4: 311–20. https://www.nature.com/articles/s41560-019-0347-9

FoE (2017) 'Switching on: How renewables will power the UK'. London: Friends of the Earth. www.foe.co.uk/sites/default/files/downloads/switching-how-renewables-will-power-uk-103641.pdf

Fowlie, M. (2019) 'The trouble with carbon pricing'. Berkeley, CA: Energy Institute at Haas, University of California, 29 April. https://energyathaas.wordpress.com/2019/04/29/the-trouble-with-carbon-pricing/

Frangoul, A. (2018) 'Here are six of China's ambitious, mind-boggling, renewable energy projects'. CNBC, 22 January. https://www.cnbc.com/2018/01/22/here-are-six-of-chinas-ambitious-mind-boggling-renewable-energy-projects.html

Fraunhofer (2016) 'Supergrid: Approach for the integration of renewable energy in Europe and North Africa'. Freiburg: Fraunhofer Institute for Solar Energy Systems. https://www.ise.fraunhofer.de/content/dam/ise/en/documents/publications/studies/Study_Supergrid_final_160412_.pdf

Fridley, D. and Heinberg, R. (2018) 'What will it take to avert collapse?'. Post Carbon Institute, 21 September. https://www.postcarbon.org/what-will-it-take-to-avert-collapse/

Gabbatiss, J. (2018) 'Africa to suffer major blackouts as climate change dries up hydropower dams, scientists warn'. *The Independent*, 30 August. http://www.independent.co.uk/environment/africa-hydropower-dams-climate-change-drought-renewable-energy-rivers-a8513331.html

Gammer, D. (2015) 'The role of hydrogen storage in a clean responsive power system'. Energy Technologies Institute. http://www.eti.co.uk/insights/carbon-capture-and-storage-the-role-of-hydrogen-storage-in-a-clean-responsive-power-system/

Garson, K. (2019) 'The bioenergy delusion'. *The Ecologist*, 21 May. https://theecologist.org/2019/may/21/bioenergy-delusion

Gautier, P. (2019) 'The inescapably slow deployment of energy technologies'. *Resilience*, 27 February. https://www.resilience.org/stories/2019-02-27/the-inescapably-slow-deployment-of-energy-technologies/

GCP (2020) Global Carbon Atlas: http://www.globalcarbonatlas.org/en/CO2-emissions

GENI (2016) '100% renewable energy reports'. San Diego, CA: Global Energy Network Institute. http://www.geni.org/globalenergy/research/100-percent-renewable-energy-reports/index.shtml

Goodall, C. (2017) 'Keeping David MacKay's "Sustainable energy – without the hot air" up-to-date'. *Carbon Commentary*, 30 March. https://www.carboncommentary.com/blog/2017/3/30/l6qcqgoedse1wmjjz87t09usoq6jva

Green, J. (2017) 'Chinese slowdown may end nuclear's last hope for growth'. *Energy Post EU*, 18 October. https://energypost.eu/chinese-slowdown-may-end-nuclears-last-hope-for-growth/

Green, M. (2019) '"New economics": The way to save the planet?'. *Reuters*, 8 May. https://www.reuters.com/article/us-climatechange-extinction/new-economics-the-way-to-save-the-planet-idUSKCN1SE2CN

Greenpeace (2015) 'Energy [r]evolution'. Greenpeace International. https://storage.googleapis.com/planet4-netherlands-stateless/2018/06/Energy-Revolution-2015-Full.pdf

Greenpeace (2019) 'How government should address the climate emergency'. Greenpeace UK. https://www.greenpeace.org.uk/wp-content/uploads/2019/04/0861_GP_ClimateEmergency_Report_Pages.pdf

Grubler, A., Wilson, C., Bento, N., et al. (2018) 'A low energy demand scenario for meeting the 1.5°C target and sustainable development goals without negative emission technologies'. *Nature Energy* 3: 515–27. https://www.nature.com/articles/s41560-018-0172-6.epdf

Grundy, A. (2019) 'Miliband brands Treasury net zero sums "slightly disturbing"'. *Current News*, 3 July. http://www.current-news. co.uk/news/miliband-brands-treasury-net-zero-cost-predictions-slightly-disturbing-in-moment-of-huge-opportunity

GWEC (2018) Global Wind Energy Council data. https://gwec.net/ global-figures/graphs/

Haas, B. (2017) 'Climate change: China calls US "selfish" after Trump seeks to bring back coal'. *The Guardian*, 30 March. https://www.theguardian.com/world/2017/mar/30/climate-change-china-us-selfish-trump-coal

Hall, C., Lambert, J. and Balogh, S. (2014) 'EROI of different fuels and the implications for society'. *Energy Policy* 64: 141–52. https://www.sciencedirect.com/science/article/pii/ S0301421513003856#bib30

Hanley, S. (2019) 'Stanford study examines hydrogen as a commercially viable storage medium for renewable energy'. *Clean Technica*, 17 July. https://cleantechnica.com/2019/07/17/ stanford-study-examines-hydrogen-as-a-storage-medium-for-renewable-energy/

Hardt, L., Owen, A., Brockway, P., et al. (2018) 'Untangling the drivers of energy reduction in the UK productive sectors: Efficiency or offshoring?'. *Applied Energy* 223: 124–33. https:// www.sciencedirect.com/science/article/pii/S0306261918304653

Harrison, R. (2017) 'Viewpoint: Why EVs aren't a silver bullet for the particulate problem'. *The Engineer*, 22 November. https:// www.theengineer.co.uk/electric-vehicles-and-particulates/

Hausfather, Z. (2018) 'Analysis: How "natural climate solutions" can reduce the need for BECCS'. *Carbon Brief*, 21 May. http:// www.carbonbrief.org/analysis-how-natural-climate-solutions-can-reduce-the-need-for-beccs

Hausfather, Z. (2019) 'Factcheck: How electric vehicles help to tackle climate change'. *Carbon Brief*, 13 May. https://www.carbonbrief. org/factcheck-how-electric-vehicles-help-to-tackle-climate-change

Heinberg, R. (2011) *The End of Growth*. Corvallis, OR: Post Carbon Institute/New Society Publishers.

Heinberg, R. (2015) 'Our renewable future'. Corvallis, OR: Post Carbon Institute. 21 January. http://www.postcarbon.org/ our-renewable-future-essay/

Heptonstall, P., Gross, R. and Steiner, F. (2017) 'The costs and impacts of intermittency – 2016 update'. UK Energy Research Centre report. http://www.ukerc.ac.uk/news/government-must-act-urgently-on-power-system-flexibility-to-avoid-costs-escalating.html

Hertwich, E., Gibon, T., Bouman, E., et al. (2014) 'Integrated life-cycle assessment of electricity-supply scenarios confirms global environmental benefit of low-carbon technologies'. *PNAS*. http:// www.pnas.org/content/early/2014/10/02/1312753111.abstract

Hickel, J. (2018) 'Why growth can't be green'. *Foreign Policy*, 12 September.   https://foreignpolicy.com/2018/09/12/why-growth-cant-be-green/

Hill, J. (2018) 'DNV GL predicts global energy demand to peak in 2035'. *CleanTechnica*, 13 September. https:// cleantechnica.com/2018/09/13/dnv-gl-predicts-global-energy-demand-to-peak-in-2035/

Hinrichs-Rahlwes, R. (2019) 'Energy policies at crossroads', in A. Sayigh (ed.), *Renewable Energy and Sustainable Buildings: Selected Papers from the World Renewable Energy Congress WREC 2018*. Charn: Springer.

Hockenos, P. (2018) 'Carbon crossroads: Can Germany revive its stalled energy transition?'. *Yale Environment* 360, 13 December. https://e360.yale.edu/features/carbon-crossroads-can-germany-revive-its-stalled-energy-transition

Hodges, J. (2019) 'Dams and reservoirs used for hydropower threaten world's rivers'. Bloomberg, 9 May. https://www. bloomberg.com/news/articles/2019-05-09/dams-and-reservoirs-used-for-hydropower-threaten-world-s-rivers

Holder, M. (2018) 'Could a power price "cannibalisation effect" put renewables investment at risk?'. *Business Green*, 24 May. https://www.businessgreen.com/bg/news-analysis/3032713/could-a-power-price-cannibalisation-effect-put-renewables-investment-at-risk

Holder, M. (2019) 'Study: Fossil fuels offer "increasingly poor" return on energy investment'. *Business Green*, 12 July. https:// www.businessgreen.com/bg/news/3078832/study-fossil-fuels-offer-increasingly-poor-return-on-energy-investment

Holliman, N. (2019) 'SoCalGas and Electrochaea announce commissioning of new biomethanation reactor system pilot project'. *Global Energy*, 14 August. https://www.energyglobal. com/bioenergy/14082019/socalgas-and-electrochaea-announce-commissioning-of-new-biomethanation-reactor-system-pilot-project/

Hornborg, A. (2019) 'A globalised solar-powered future is wholly unrealistic – and our economy is the reason why'. *The Conversation*, 6 September. https://theconversation. com/a-globalised-solar-powered-future-is-wholly-unrealistic-and-our-economy-is-the-reason-why-118927

Hu, J., Harmsen, R., Crijns-Graus, W. and Worrell, E. (2019) 'Geographical optimization of variable renewable energy capacity in China using modern portfolio theory'. *Applied Energy*

253.   https://www.researchgate.net/publication/335014153_ Geographical_optimization_of_variable_renewable_energy_ capacity_in_China_using_modern_portfolio_theory

Huenteler, J., Tang, T., Chan, G. and Diaz Anadon, L. (2018) 'Why is China's wind power generation not living up to its potential?'. *Environmental Research Letters* 13(4). http://iopscience.iop.org/ article/10.1088/1748-9326/aaadeb

IAEA (2019) 'Nuclear hydrogen production'. Vienna: International Atomic Energy Agency. https://www.iaea.org/topics/non- electric-applications/nuclear-hydrogen-production

ICAX (2019) 'Interseasonal heat stores'. Company web site. http:// www.icax.co.uk/interseasonal_heat_transfer.html

ICL (2018) 'Analysis of alternative UK heat decarbonisation pathways'. Imperial College London report for the Committee on Climate Change. http://www.theccc.org.uk/publication/ analysis-of-alternative-uk-heat-decarbonisation-pathways

ICL/Ovo (2018) 'Blueprint for a post-carbon society'. Imperial College London/Ovo Energy. https://www.ovoenergy.com/ binaries/content/assets/documents/pdfs/newsroom/blueprintfora postcarbonsociety-2018.pdf

IEA (2015) 'Carbon capture and storage'. Paris: International Energy Agency. https://www.iea.org/reports/ carbon-capture-and-storage-2015

IEA (2017) 'IEA bioenergy response to Chatham House report "Woody biomass for power and heat: impacts on the global climate"'. Paris: IEA Bioenergy, International Energy Association. http://www.ieabioenergy.com/publications/iea- bioenergy-response/

IEA (2018a) 'Renewables 2018: Analysis and forecasts to 2023'. Paris: International Energy Association. https://www.iea.org/ renewables2018/power/

IEA (2018b) 'Energy efficiency 2018: Analysis and outlooks to 2040'. Paris: International Energy Association. https://www.iea. org/efficiency2018/

IEA (2019a) 'World energy outlook'. Paris: International Energy Agency. https://www.iea.org/weo/

IEA (2019b) 'World energy investment'. Paris: International Energy Agency. https://webstore.iea.org/world-energy-investment-2019

IEA (2019c) 'The Future of Hydrogen'. Report for the 2019 G20 summit, Japan. Paris: International Energy Agency. http://www. iea.org/reports/the-future-of-hydrogen

IEA (2019d) 'Tracking Transport: Aviation'. A tracking clean energy progress briefing. Paris: International Energy Agency. https://www.iea.org/tcep/transport/aviation/

IEA (2019e) 'Steep decline in nuclear power would threaten energy security and climate goals'. Paris: International Energy Agency. www.iea.org/newsroom/news/2019/may/steep-decline-in-nuclear-power-would-threaten-energy-security-and-climate-goals.html

IEA (2019f) 'Securing investments in low-carbon power generation sources'. G20 submission from the International Energy Agency, Paris. https://webstore.iea.org/securing-investments-in-low-carbon-power-generation-sources

IEA (2019g) 'Renewables 2019'. Paris: International Energy Agency. https:www.iea.org/reports/renewables-2019

IEA (2020) 'Global CO_2 emissions in 2019'. Paris: International Energy Agency. http://www.iea.org/articles/global-co2-emissions-in-2019

IET (2019) 'Transitioning to hydrogen'. London: Institution of Engineering and Technology. http://www.theiet.org/media/4095/transitioning-to-hydrogen.pdf

IMF (2019) 'World economic outlook'. Washington, DC: International Monetary Fund. http://www.imf.org/en/Publications/WEO/Issues/2019/03/28/world-economic-outlook-april-2019

IPBES (2019) 'Global assessment report on biodiversity and ecosystem services'. UN Intergovernmental Science-Policy Platform on Biodiversity and Ecosystem Services. https://www.ipbes.net/global-assessment-report-biodiversity-ecosystem-services

IPCC (2019) Ongoing series of Assessment Reports and Special Reports. Geneva: Intergovernmental Panel on Climate Change. https://www.ipcc.ch/library/

IPPR (2018) 'A distributed energy future for the UK'. London: Institute for Public Policy Research. http://www.ippr.org/files/2018-09/a-distributed-energy-future-for-the-uk-september18.pdf

IRENA (2017a) 'Perspectives for the energy transition: Investment needs for a low-carbon energy system'. Abu Dhabi: International Renewable Energy Agency, joint report with the International Energy Agency, Paris. https://www.irena.org/publications/2017/Mar/Perspectives-for-the-energy-transition-Investment-needs-for-a-low-carbon-energy-system

IRENA (2017b) 'Geothermal power technology brief'. Abu Dhabi: International Renewable Energy Agency. http://www.irena.org/publications/2017/Aug/Geothermal-power-Technology-brief

IRENA (2017c) 'Electric vehicles: Technology brief'. Abu Dhabi: International Renewable Energy Agency. http://www.irena.org/DocumentDownloads/Publications/IRENA_Electric_Vehicles_2017.pdf

IRENA (2018) 'Global energy transformation: A roadmap to 2050'.

Abu Dhabi: International Renewable Energy Agency. https://irena.org/publications/2018/Apr/Global-Energy-Transition-A-Roadmap-to-2050

IRENA (2019a) 'Wind energy'. Abu Dhabi: International Renewable Energy Agency. https://www.irena.org/wind

IRENA (2019b) 'Hydrogen: A renewable energy perspective'. Abu Dhabi: International Renewable Energy Agency. http://www.irena.org/-/media/Files/IRENA/Agency/Publication/2019/Sep/IRENA_Hydrogen_2019.pdf

IRENA (2019c) 'A new world: The geopolitics of the energy transformation'. Abu Dhabi: Global Commission on the Geopolitics of Energy Transformation, International Renewable Energy Agency. http://geopoliticsofrenewables.org/assets/geopolitics/Reports/wp-content/uploads/2019/01/Global_commission_renewable_energy_2019.pdf

IRENA (2019d) 'Solutions to integrate high shares of variable renewable energy'. Submission to the G20 summit from International Renewable Energy Agency, Abu Dhabi. https://www.irena.org/publications/2019/Jun/Solutions-to-integrate-high-shares-of-variable-renewable-energy

IRENA (2019e) 'The future of wind'. Abu Dhabi: International Renewable Energy Agency. http://www.irena.org/publications/2019/Oct/Future-of-wind

Irfan, U. (2019) 'After years of decline, US carbon emissions are rising again'. *Vox*, 9 January. https://www.vox.com/2019/1/8/18174082/us-carbon-emissions-2018

ITA (2016) '2016 top markets report renewable energy'. Washington, DC: US International Trade Administration. https://legacy.trade.gov/topmarkets/pdf/Renewable_Energy_Executive_Summary.pdf

Jackson, T. (2009) *Prosperity without Growth*. Sustainable Development Commission. London: Routledge.

Jackson, T. (2018) 'How the light gets in'. CUSP, 4 November. http://www.cusp.ac.uk/themes/aetw/blog_tj_how-the-light-gets-in/

Jacobson, L. (2018) 'Are greenhouse emissions down under Donald Trump, as EPA says?'. PolitiFact, 18 June. http://www.politifact.com/truth-o-meter/statements/2018/jun/18/environmental-protection-agency/are-greenhouse-emissions-down-under-donald-trump-e/

Jacobson, M. (2019a) '100% clean, renewable energy and storage for everything'. Updated data extract from forthcoming Cambridge University Press book, 13 January. https://web.stanford.edu/group/efmh/jacobson/Articles/I/TimelineDetailed.pdf

Jacobson, M. (2019b) 'Guide to the 139 country 100%

renewables scenario'. https://ars.els-cdn.com/content/image/1-s2.0-S2542435117300120-mmc2.mp4

Jacobson, M. (2019c) 'Why not biomass for electricity or heat as part of a 100% wind–water–solar (WWS) and storage solution to global warming, air pollution, and energy security'. Pre-publication extract from forthcoming Cambridge University Press book, 23 July. https://web.stanford.edu/group/efmh/jacobson/Articles/I/BiomassVsWWS.pdf

Jacobson, M. and Delucchi, M. (2009) 'A plan to power 100 percent of the planet with renewables'. *Scientific American*, November. http://www.scientificamerican.com/article.cfm?id=a-path-to-sustainable-energy-by-2030

Jacobson, M. and Delucchi, M. (2011) 'Providing all global energy with wind, water, and solar power, Part I: Technologies, energy resources, quantities and areas of infrastructure, and materials'. *Energy Policy* 39(3): 1154–69. https://www.sciencedirect.com/science/article/pii/S0301421510008645

Jacobson, M., Delucchi, M., Bauer, Z., et al. (2017) '100% clean and renewable wind, water, and sunlight all-sector energy roadmaps for 139 countries of the world'. *Joule* 1(1): 108–21. http://www.sciencedirect.com/science/article/pii/S2542435117300120

Jacobson, M., Delucchi, M., Bazouin, G., et al. (2015) '100% clean and renewable wind, water, and sunlight (WWS) all-sector energy roadmaps for the 50 United States'. *Energy and Environmental Science* 8, 2093. http://web.stanford.edu/group/efmh/jacobson/Articles/I/USStatesWWS.pdf

Japan Times (2019) 'Full text of the G20 Osaka leaders' declaration', including US clause 36, 29 June. https://www.japantimes.co.jp/news/2019/06/29/national/full-text-g20-osaka-leaders-declaration/

Järvensivu, P., Toivanen, T., Vadén, T., Lähde, V., Majava, A. and Eronen, J. (2018) Global Sustainable Development report by a group of Finnish academics. https://bios.fi/bios-governance_of_economic_transition.pdf

Johnson, N. (2018) 'The debate is over: we need to start sucking carbon from the air'. *Grist*, 24 October. https://grist.org/article/the-debate-is-over-we-need-to-start-sucking-carbon-from-the-air/

Johnson, S. (2018) 'Scientists to UN: To stop climate change, modern capitalism needs to die'. *Big Think*, 29 August. https://bigthink.com/stephen-johnson/scientists-to-un-to-stop-climate-change-modern-capitalism-needs-to-die

JRC (2016) 'JRC ocean energy status report 2016 edition'. European

Commission Joint Research Centre. https://setis.ec.europa.eu/sites/default/files/reports/ocean_energy_report_2016.pdf

Katwala, A. (2018) 'The spiralling environmental cost of our lithium battery addiction'. *Wired*, 5 August. https://www.wired.co.uk/article/lithium-batteries-environment-impact

Kaufman, R. (2019) 'Self-driving cars have every incentive to create havoc'. *Next City*, 6 February. https://nextcity.org/daily/entry/self-driving-cars-have-every-incentive-to-create-havoc

Kazaglis, A., Aaron Tam, A., Eis, J., et al. (2019) 'Accelerating innovation towards net zero emissions'. Report for the Aldersgate Group by UKERC/Vivid Economics. http://www.ukerc.ac.uk/publications/aldersgate-report-net-zero.html

Keating, D. (2019) 'EU decarbonization plan for 2050 collapses after Polish veto'. *Forbes*, 20 June. http://www.forbes.com/sites/davekeating/2019/06/20/eu-decarbonisation-plan-for-2050-collapses-after-polish-veto

Kidd, S. (2017) 'Nuclear in China – why the slowdown?'. *Nuclear Engineering International*, 10 August. https://www.neimagazine.com/opinion/opinionnuclear-in-china-why-the-slowdown-5896525/

Kirschbaum, E. (2019) 'Germany to close all 84 of its coal-fired power plants, will rely primarily on renewable energy'. *Los Angeles Times*, 26 January. https://www.latimes.com/world/europe/la-fg-germany-coal-power-20190126-story.html

Klebnikov, S. (2019) 'Stopping global warming will cost $50 trillion: Morgan Stanley Report'. *Forbes*, 24 October. https://www.forbes.com/sites/sergeiklebnikov/2019/10/24/stopping-global-warming-will-cost-50-trillion-morgan-stanley-report/#122b851651e2

Koirala, P. B., Koliou, E., Friege, J., Hakvoort, R. and Herde, P. (2016) 'Energetic communities for community energy: A review of key issues and trends shaping integrated community energy systems'. *Renewable and Sustainable Energy Reviews*, 56: 722–44. http://www.sciencedirect.com/science/article/pii/S1364032115013477

Konoplyanik, A. (2019) 'On the new paradigm of international energy development: Risks and challenges for Russia and the world on the way to the low-carbon future', in G. Wood and K. Baker (eds), *The Palgrave Handbook of Managing Fossil Fuels and Energy Transitions*. Basingstoke: Palgrave.

Kroposki, B. (2016) 'Can smarter solar inverters save the grid?'. *IEEE Spectrum*, 20 October. http://spectrum.ieee.org/energy/renewables/can-smarter-solar-inverters-save-the-grid

Labour (2019a) 'Bringing energy home'. UK Labour Party

policy document. http://www.labour.org.uk/wp-content/uploads/2019/03/Bringing-Energy-Home-2019.pdf

Labour (2019b) 'Thirty recommendations by 2030'. Expert briefing for the Labour Party. https://labour.org.uk/wp-content/uploads/2019/10/ThirtyBy2030report.pdf

Labour GND (2019) UK Labour Green New Deal outline. https://www.labourgnd.uk/gnd-explained

Lambert. M. (2018) 'Power-to-gas: Linking electricity and gas in a decarbonising world?'. Oxford Institute for Energy Studies. http://www.oxfordenergy.org/wpcms/wp-content/uploads/2018/10/Power-to-Gas-Linking-Electricity-and-Gas-in-a-Decarbonising-World-Insight-39.pdf

Lance, N., De Rubens, G., Kester, J. and Sovocool, B. (2019) *Vehicle-to-Grid: A Sociotechnical Transition beyond Electric Mobility*. Basingstoke: Palgrave.

Lazard (2018a) 'Levelized cost of energy and levelized cost of storage 2018'. Lazard consultants. http://www.lazard.com/perspective/levelized-cost-of-energy-and-levelized-cost-of-storage-2018/

Lazard (2018b) 'Lazard's levelized cost of energy analysis'. Lazard consultants. http://www.lazard.com/media/450773/lazards-levelized-cost-of-energy-version-120-vfinal.pdf

Lee, A. (2019) 'Floating wind-to-hydrogen plan to heat millions of UK homes'. *Recharge News*, 13 September. https://www.rechargenews.com/wind/1850034/floating-wind-to-hydrogen-plan-to-heat-millions-of-uk-homes

Lib Dems (2019) 'Tackling the climate emergency'. Liberal Democrat Policy Paper 139. https://d3n8a8pro7vhmx.cloudfront.net/libdems/pages/46346/attachments/original/1564404765/139_-_Tackling_the_Climate_Emergency_web.pdf?1564404765

Liebreich, M. (2018) 'The secret of eternal growth'. Initiative for Free Trade, 29 October. http://ifreetrade.org/article/the_secret_of_eternal_growth_the_physics_behind_pro_growth_environmentalism

Liebreich, M. (2019) 'We need to talk about nuclear power'. Bloomberg New Energy Finance blog, 3 July. https://about.bnef.com/blog/liebreich-need-talk-nuclear-power/

Liu, H. (2016) 'The dark side of renewable energy'. *Earth Journalism Network*, 25 August. https://earthjournalism.net/stories/the-dark-side-of-renewable-energy

Liu, Y. (2018) 'Wind power curtailment in China on the mend'. *Renewable Energy World*, 26 January. https://www.renewableenergyworld.com/articles/2018/01/wind-power-curtailment-in-china-on-the-mend.html

Liu, Z. (2016) *Global Energy Interconnection*. New York/

Amsterdam: Academic Press. https://www.sciencedirect.com/book/9780128044056/global-energy-interconnection#book-description

Lomborg, B. (2019) 'No, renewables are not taking over the world'. Facebook post and video, 30 March. https://www.facebook.com/bjornlomborg/videos/no-renewables-are-not-taking-over-the-worldwere-constantly-being-told-how-renewa/391717398051053/

Lombrana, L. M. (2019) 'Saving the planet with electric cars means strangling this desert'. Bloomberg, 11 June. https://www.bloomberg.com/news/features/2019-06-11/saving-the-planet-with-electric-cars-means-strangling-this-desert

Londe, L. (2018) 'Hydrogen caverns are a proven, inexpensive and reliable technology'. Medium Corporation, 8 February. https://medium.com/@cH2ange/louis-londe-technical-director-at-geostock-hydrogen-caverns-are-a-proven-inexpensive-and-346dde79c460

Lovegrove, P. (2018) 'The world's energy demand will peak in 2035 prompting a reshaping of energy investment'. DNV-GL, 10 September. https://www.dnvgl.com/news/the-world-s-energy-demand-will-peak-in-2035-prompting-a-reshaping-of-energy-investment-128751

Lovins, A. (2018) 'How big is the energy efficiency resource?'. *Environmental Research Letters* 13(9). http://iopscience.iop.org/article/10.1088/1748-9326/aad965

Lovins, A. and Nanavatty, R. (2019) 'A market-driven Green New Deal? We'd be unstoppable'. *New York Times*, 18 April. http://www.nytimes.com/2019/04/18/opinion/green-new-deal-climate.html

Lovins, A., Palazzi, T., Laemel, R. and Goldfield, E. (2018) 'Relative deployment rates of renewable and nuclear power: A cautionary tale of two metrics'. *Energy Research & Social Science* 38: 188–92. https://www.sciencedirect.com/science/article/pii/S2214629618300598?via%3Dihub

Luo, X., Wang, J., Dooner, M., Clarke, J. and Krupke, C. (2014) 'Overview of current development in compressed air energy storage technology'. *Energy Procedia* 62: 603–11. https://www.sciencedirect.com/science/article/pii/S1876610214034547#!

Lyman, R. (2019) 'Transition to reality: The prospects for rapid global decarbonization'. London: Global Warming Policy Foundation. https://www.thegwpf.org/content/uploads/2019/02/Lyman-2019.pdf

MacDonald, A., Clack, C., Alexander, A., Dunbar, A., Wilczak, J. and Xie, Y. (2016) 'Future cost-competitive electricity systems and their impact on US CO_2 emissions'. *Nature*

Climate Change 6: 526–31, online. http://www.researchgate. net/publication/291525923_Future_cost-competitive_electricity_ systems_and_their_impact_on_US_CO2_omissions

McCauley, D. (2019) *Energy Justice: Re-balancing the Trilemma of Security, Poverty and Climate Change*. Basingstoke: Palgrave.

Mace, M. (2019) 'Dutch launch prototype solar car with 720km range'. *Euractiv*, 2 June. https://www.euractiv.com/section/electric-cars/ news/dutch-launch-prototype-solar-car-with-720km-range/

McGee, J. A. and Greiner, P. T. (2019) 'Renewable energy injustice: The socio-environmental implications of renewable energy consumption'. *Energy Research & Social Science* 56: 101214. http://www.sciencedirect.com/science/article/pii/S22146 29618310971?via%3Dihub

MacKay, D. (2013) 'Sustainable energy – without the hot air'. Available as an online book, with some updates. http://www. withouthotair.com/

McLaren, D. (2019) 'The problem with net-zero emissions targets'. *Carbon Brief*, 30 September. https://www.carbonbrief.org/ guest-post-the-problem-with-net-zero-emissions-targets

McMahon, J. (2019) 'How the clean energy transition could save more than it costs'. *Forbes*, 5 August. http://www. forbes.com/sites/jeffmcmahon/2019/08/05/how-the-clean- energy-transition-could-save-more-than-it-costs

McPhee, D. (2017) 'The top five mega solar projects in China'. Energy Voice video. https://www.energyvoice.com/ otherenergy/157332/watch-top-five-mega-solar-projects-china/

Mahajan, M. (2018) 'Plunging prices mean building new renewable energy is cheaper than running existing coal'. *Forbes*, 3 December. https://www.forbes.com/sites/energyinnovation/2018/12/03/ plunging-prices-mean-building-new-renewable-energy-is- cheaper-than-running-existing-coal/#318d78c531f3

Marinucci, C. and Kahn, D. (2019) 'Labor anger over Green New Deal greets 2020 contenders in California'. *Politico*, 6 June. https://www.politico.com/states/california/story/2019/06/01/ labor-anger-over-green-new-deal-greets-2020-contenders-in- california-1027570

Martin, I. (2019) 'Democracy and progress are facing extinction'. *The Times*, 19 July.

Marvel, K., Kravitz, B. and Caldeira, K. (2013) 'Geophysical limits to global wind power'. *Nature Climate Change* 3: 118–21. https://www.nature.com/articles/nclimate1683

May, A. (2019) 'World's energy transition in doubt as progress on affordability, sustainability stalls'. World Economic Forum blog, 25 March. https://www.weforum.org/press/2019/03/

world-s-energy-transition-in-doubt-as-progress-on-affordability-sustainability-stalls/

Merchant, E. F. (2018) 'China's bombshell solar policy shift could cut expected capacity by 20 gigawatts'. Greentech Media, 6 June. https://www.greentechmedia.com/articles/read/chinas-bombshell-solar-policy-could-cut-capacity-20-gigawatts#gs.tqawln

Merchant, E. F. (2019) 'DNV GL: Renewables at 80% of global electricity mix by 2050'. Greentech Media, 11 September. https://www.greentechmedia.com/articles/read/dnv-gl-renewables-at-80-of-global-electricity-mix-by-2050#gs.2spze5

Mills, M. (2019) 'Inconvenient energy realities'. *Economics* 21, 1 July. https://economics21.org/inconvenient-realities-new-energy-economy

Millward, D. (2018) 'Future is dim for US nuclear power plants'. *The Nation*, 17 December. http://www.thenational.ae/world/the-americas/future-is-dim-for-us-nuclear-power-plants-1.803434

Mitchell, C., Hoggett, R., Willis, B. and Britton, J. (2019) 'Submission to Ofgem targeted charging review: Minded to decision and draft impact assessment'. Energy Policy Group, University of Exeter. http://projects.exeter.ac.uk/igov/wp-content/uploads/2019/02/Exeter-EPG-response-to-Ofgem-Targeted-Charging-Review-Feb-2019.pdf

Monbiot, G. (2019) 'Dare to declare capitalism dead – before it takes us all down with it'. *The Guardian*, 25 April. https://www.theguardian.com/commentisfree/2019/apr/25/capitalism-economic-system-survival-earth

Montague, B. (2018) 'Hardwood forests cut down to feed Drax Power plant, Channel 4 Dispatches claims'. *The Ecologist*, 16 April. https://theecologist.org/2018/apr/16/hardwood-forests-cut-down-feed-drax-power-plant-channel-4-dispatches-claims

Montford, A. (2019) 'Green killing machines'. London: Global Warming Policy Foundation. http://www.thegwpf.org/content/uploads/2019/07/Green-Killing-Machines-1.pdf

Monyei, M., Sovacool, B., Brown, M., et al. (2018) 'Justice, poverty, and electricity decarbonization'. *Electricity Journal* 32(1): 47–51. https://www.sciencedirect.com/science/article/pii/S1040619018303142?via%3Dihub

Moran, E., Lopez, M. C., Moore, N., Müller, N. and Hyndman, D. (2018) 'Sustainable hydropower in the 21st century'. *PNAS*, 5 November. http://www.pnas.org/content/early/2018/11/02/1809426115

Morgan, S. (2018) 'EU study weighs linking power grid to China's'. *Euractiv*, 21 March. https://www.euractiv.com/section/eu-china/news/eu-looks-into-benefits-of-energy-silk-road-to-china/

Morgan, S. (2019) 'The brief – Poland's failure to think big'. *Euractiv*, 24 June. http://www.euractiv.com/section/energy-environment/news/the-brief-polands-failure-to-think-big

Morris, C. (2018) 'Can reactors react?. IASS Discussion Paper. https://www.iass-potsdam.de/en/output/publications/2018/can-reactors-react-decarbonized-electricity-system-mix-fluctuating

Mouli-Castillo, J., Wilkinson, M., Mignard, D., McDermott, C., Haszeldine, S. and Shipton, Z. (2019) 'Inter-seasonal compressed-air energy storage using saline aquifers'. *Nature Energy*, 21 January. https://www.nature.com/articles/s41560-018-0311-0

Movellan, J. (2016) 'The Asia super grid – Four countries join together to maximize renewable energy'. *Renewable Energy World*, 18 October. https://www.renewableenergyworld.com/articles/2016/10/the-asia-super-grid-countries-join-together-to-maximize-renewable-energy.html

Muffett, C. (2019) 'Geoengineering is a dangerous distraction'. Heinrich Böll Stiftung. https://www.boell.de/en/2019/02/18/geoengineering-dangerous-distraction

Murray, J. (2019) 'Net zero: Battalion of corporate giants pledge to deliver 1.5C climate targets'. *Business Green*, 23 July. http://www.businessgreen.com/bg/news/3079369/net-zero-battalion-of-corporate-giants-pledge-to-deliver-15c-climate-targets

Myhrvold, N. (2018) 'Why we need innovative nuclear power'. *Scientific American*, 7 November. https://blogs.scientificamerican.com/observations/why-we-need-innovative-nuclear-power/

Myllyvirta, L. (2019) 'India's CO2 emissions growth poised to slow sharply in 2019'. *Carbon Brief*, 24 October. https://www.carbonbrief.org/analysis-indias-co2-emissions-growth-poised-to-slow-sharply-in-2019

Navigant (2019) 'Gas for climate'. Navigant consultancy, the Netherlands. https://www.gasforclimate2050.eu/files/files/Navigant_Gas_for_Climate_The_optimal_role_for_gas_in_a_net_zero_emissions_energy_system_March_2019.pdf

Nazari, A. and Sanjayan, J. (2017) *Handbook of Low Carbon Concrete*. London: Elsevier.

Neslen, A. (2018) 'Spain plans switch to 100% renewable electricity by 2050'. *The Guardian*, 13 November. https://www.theguardian.com/environment/2018/nov/13/spain-plans-switch-100-renewable-electricity-2050

NEU (2018) 'Middle East and North Africa wind capacity forecast to hit 23 GW by 2027'. New Energy Update, 23 May. http://analysis.newenergyupdate.com/wind-energy-update/middle-east-and-north-africa-wind-capacity-forecast-hit-23-gw-2027

Newell, R., Raimi, D. and Aldana, G. (2019) 'Global Energy Outlook 2019: The next generation of energy'. Washington, DC: Resources for the Future. https://www.rff.org/publications/reports/global-energy-outlook-2019/

Newman, A. (2018) 'Scientists ridicule latest round of federal "climate change" hysteria'. *New American*, 26 November. https://www.thenewamerican.com/tech/environment/item/30735-scientists-ridicule-latest-round-of-federal-climate-change-hysteria

New Power (2019) 'Offshore wind hits record low below £40/MWh'. *New Power*, 20 September. http://www.newpower.info/2019/09/offshore-wind-hits-record-low-below-40mwh/

NG (2018) 'Future energy scenarios'. National Grid. http://fes.nationalgrid.com/media/1363/fes-interactive-version-final.pdf

Ng, E. (2017) 'China to erect fewer farms, generate less solar power in 2017'. *South China Morning Post*, 19 April. https://www.scmp.com/business/article/2088905/china-erect-fewer-farms-generate-less-solar-power-2017

NHM (2019) 'Leading scientists set out resource challenge of meeting net zero emissions in the UK by 2050'. London: Natural History Museum, 5 June. www.nhm.ac.uk/press-office/press-releases/leading-scientists-set-out-resource-challenge-of-meeting-net-zer.html

NIC (2017) 'Smart power'. London: National Infrastructure Commission. http://www.gov.uk/government/uploads/system/uploads/attachment_data/file/505218/IC_Energy_Report_web.pdf

NRDC (2016) 'Money to burn? The UK needs to dump biomass and replace its coal plants with truly clean energy'. Washington, DC: Natural Resources Defense Council. https://www.nrdc.org/sites/default/files/uk-biomass-replace-coal-clean-energy-ib.pdf

NREL (2012) 'Renewable electricity futures study'. Golden, CO: National Renewable Energy Laboratory. https://www.nrel.gov/analysis/re-futures.html

Obama, B. (2009) Presidential speech to Congress. http://www.whitehouse.gov/the_press_office/Remarks-of-President-Barack-Obama-Address-to-Joint-Session-of-Congress

OECD (2018) 'The costs of decarbonisation: System costs with high shares of nuclear and renewables'. Paris: Nuclear Energy Agency/OECD. https://www.oecd-nea.org/ndd/pubs/2019/7299-system-costs.pdf

O'Neil, B. (2018) 'In praise of the gilets jaunes'. *The Spectator*, 3 December. https://blogs.spectator.co.uk/2018/12/in-praise-of-the-gilets-jaunes/

Ørsted (2017) 'Green energy barometer'. Edelman Intelligence for Ørsted. https://orsted.com/barometer

Overland, I. (2019) 'The geopolitics of renewable energy: Debunking four emerging myths'. *Energy Research & Social Science* 49: 36–40. https://www.sciencedirect.com/science/article/pii/S2214629618308636

Ovo/Imperial (2018) 'Blueprint for a post-carbon society'. Ovo Energy/Imperial College London. http://www.ovoenergy.com/binaries/content/assets/documents/pdfs/newsroom/blueprint-for-a-post-carbon-society-how-residential-flexibility-is-key-to-decarbonising-power-heat-and-transport/blueprintforapostcarbonsocietypdf-compressed.pdf

Parrique, T., Barth, J., Briens, F., Kerschner, C., Kraus-Polk A., Kuokkanen, A. and Spangenberg, J. H. (2019) 'Decoupling debunked'. European Environmental Bureau. https://eeb.org/library/decoupling-debunked/

Partington, R. (2019a) 'To ensure a green future the UK cannot rely on free markets alone'. *The Guardian*, 7 July. https://www.theguardian.com/environment/2019/jul/07/to-ensure-a-green-future-the-uk-needs-to-ditch-free-market-economics

Partington, R. (2019b) 'Britain now G7's biggest net importer of CO2 emissions per capita, says ONS'. *The Guardian*, 21 October. https://www.theguardian.com/uk-news/2019/oct/21/britain-is-g7s-biggest-net-importer-of-co2-emissions-per-capita-says-ons

Patel, S. (2019) 'Wind industry prepares for massive expansion'. *Power Magazine*, 4 November. https://www.powermag.com/wind-industry-prepares-for-massive-expansion/

Pearce, F. (2017) 'Industry meltdown: Is the era of nuclear power coming to an end?'. *Yale Environment* 360 (15 May). http://e360.yale.edu/features/industry-meltdown-is-era-of-nuclear-power-coming-to-an-end

PEI (2018) 'Solar and wind now the cheapest power source, says Bloomberg NEF', Power Engineering International, 19 November. https://www.powerengineeringint.com/renewables/solar/solar-and-wind-now-the-cheapest-power-source-says-bloombergnef

Penn, I. (2018) 'The $3 billion plan to turn Hoover Dam into a giant battery'. *New York Times*, 24 July. http://www.nytimes.com/interactive/2018/07/24/business/energy-environment/hoover-dam-renewable-energy.html

Perez, R. and Rabago, K. (2019) 'A radical idea to get a high-renewable electric grid: Build way more solar and wind than needed'. *The Conversation*, 29 May. https://theconversation.com/a-radical-idea-to-get-a-high-renewable-electric-grid-build-way-more-solar-and-wind-than-needed-113635

PlanEnergi (2019) 'Solar district heating in Denmark

1988–2018'. http://planenergi.eu/activities/district-heating/solar-district-heating/sdh-in-dk-1988-2018/

Platts (2018) 'German 2017 emissions down 0.5% as energy drop offset by industry rise'. S&P Global, 28 March. https://www.spglobal.com/platts/en/market-insights/latest-news/electric-power/032818-german-2017-emissions-down-05-as-energy-drop-offset-by-industry-rise

Pollin, R. (2018) 'De-growth vs a Green New Deal'. *New Left Review* 112, July/August. https://newleftreview.org/issues/II112/articles/robert-pollin-de-growth-vs-a-green-new-deal

Portugal-Pereira, J., Ferreira, P., Cunha, J., Szklo, A., Schaeffer, R. and Araújo, M. (2018) 'Better late than never, but never late is better: Risk assessment of nuclear power construction projects'. *Energy Policy* 120: 158–66. https://www.sciencedirect.com/science/article/pii/S0301421518303446#!

Probert, A. (2018) 'China leads world's renewable energy investment'. *Greener Ideal*, 17 January. https://greenerideal.com/news/china-leads-worlds-renewable-energy-investment/

Pugwash (2013) 'Pathways to 2050: Three possible UK energy strategies scenario'. British Pugwash report. https://britishpugwash.org/pathways-to-2050-three-possible-uk-energy-strategies/

Purvins, A., Sereno, L., Ardelean, M., Covrig, C-F., Efthimiadis, T. and Minnebo, P. (2018) 'Submarine power cable between Europe and North America: A techno-economic analysis'. *Journal of Cleaner Production* 186: 131–45. https://www.sciencedirect.com/science/article/pii/S0959652618307522

Ram, M., Bogdanov, D., Aghahosseini, A., et al. (2019) 'Global energy system based on 100% renewable energy – power, heat, transport and desalination sectors'. Berlin: Lappeenranta University of Technology and Energy Watch Group. http://energywatchgroup.org/wp-content/uploads/EWG_LUT_100RE_All_Sectors_Global_Report_2019.pdf

Rao, C. N. R. and Dey, S. (2017) 'Solar thermochemical splitting of water to generate hydrogen'. *PNAS* Special Feature: Perspective. http://www.pnas.org/content/pnas/early/2017/05/17/1700104114.full.pdf

Rapier, R. (2019) 'Renewables catching nuclear power in global energy race'. *Forbes*, 7 July. http://www.forbes.com/sites/rrapier/2019/07/07/wind-and-solar-power-nearly-matched-nuclear-power-in-2018

Raugei, M. (2013) 'Comments on "energy intensities, EROIs (energy returned on invested), and energy payback times of electricity generating power plants" – Making clear of quite some confusion'. *Energy* 59: 781–2. http://www.sciencedirect.com/science/article/pii/S0360544213006373?np=y – aff1

RE100 (2019) RE100 web site. http://there100.org/companies

REA (2019) 'Bioenergy is the "little-known leader" in British renewables'. London: Renewable Energy Association. https://www.r-e-a.net/news/rea-bioenergy-is-the-little-known-leader-in-british-renewables

REN21 (2018) 'Renewables 2018 Global Status Report'. Paris: Renewable Energy Network for the 21st Century. http://www.ren21.net/status-of-renewables/global-status-report/

REN21 (2019) 'Renewables 2019 Global Status Report'. Paris: Renewable Energy Network for the 21st Century. http://www.ren21.net/status-of-renewables/global-status-report/

Reuters (2017) 'China energy regulator raises targets for curbing coal-fired power'. *Reuters*, 2 August. https://www.reuters.com/article/us-china-coal-utilities-idUSKBN1AI0MO

Reuters (2019a) 'China launches subsidy-free solar, wind power after project costs fall. *Reuters*, 9 January: https://uk.reuters.com/article/us-china-energy-renewables/china-launches-subsidy-free-solar-wind-power-after-project-costs-fall-idUKKCN1P30ZQ

Reuters (2019b) 'China's 2018 renewable power capacity up 12 percent on year'. *Reuters*, 28 January. https://uk.reuters.com/article/us-china-renewables/chinas-2018-renewable-power-capacity-up-12-percent-on-year-idUKKCN1PM0HM

Richard, C. (2018a) 'Cost of wind-generated hydrogen to fall below natural gas'. *Windpower Monthly*, 24 April. https://www.windpowermonthly.com/article/1462904/cost-wind-generated-hydrogen-fall-below-natural-gas

Richard, C. (2018b) 'Government offers 700MW subsidy-free'. *Windpower Monthly*, 4 December. https://www.windpowermonthly.com/article/1520238/government-offers-700mw-subsidy-free

Ritchie, H. (2017) 'What was the death toll from Chernobyl and Fukushima?'. Our World in Data. https://ourworldindata.org/what-was-the-death-toll-from-chernobyl-and-fukushima

Roberts, D. (2018) 'After rising for 100 years, electricity demand is flat: Utilities are freaking out'. *Vox*, 27 February. http://www.vox.com/energy-and-environment/2018/2/27/17052488/electricity-demand-utilities

Rogers, N. (2018) *Dumb Energy: A Critique of Wind and Solar Energy*. Self-published book. http://www.dumbenergy.com/

Romm, J. (2019) 'A 100% renewable grid isn't just feasible, it's in the works in Europe'. ThinkProgress, 21 June. https://

thinkprogress.org/europe-will-be-90-renewable-powered-in-two-decades-experts-say-8db3e7190bb7/

Roselund, C. (2019) 'Green New Deal secures support from its first national-level US labor union'. *PV Magazine*, 7 June. http://www.pv-magazine.com/2019/06/07/green-new-deal-secures-support-from-its-first-national-level-us-labor-union/

Rosenow, J., Guertler, P., Sorrell, S. and Eyre, N. (2018) 'The remaining potential for energy savings in UK households'. *Energy Policy* 121: 542–52. https://www.sciencedirect.com/science/article/pii/S030142151830421X

Rosenzweig, R. (2016) *Global Climate Change Policy and Carbon Markets*. Basingstoke: Palgrave.

RSPB (2016) 'The RSPB's 2050 energy vision'. Royal Society for the Protection of Birds. http://ww2.rspb.org.uk/Images/energy_vision_summary_report_tcm9-419580.pdf

RTP (2015) 'Distributing power: A transition to a civic energy future'. Realising Transition Pathways Research Consortium. http://www.realisingtransitionpathways.org.uk/realisingtransitionpathways/news/distributing_power.html

Russell, C. (2019) 'Coal going from winner to loser in India's energy future'. *Reuters*, 20 February. https://www.reuters.com/article/us-column-russell-coal-india/coal-going-from-winner-to-loser-in-indias-energy-future-russell-idUSKCN1Q90OP

Saltmarsh, C. (2019) 'An internationalist Green New Deal'. *The Ecologist*, 23 April. https://theecologist.org/2019/apr/23/internationalist-green-new-deal

Sanders, B. (2019) 'The Green New Deal'. Campaign blog. https://berniesanders.com/issues/the-green-new-deal

Sauar, E. (2017) 'IEA underreports contribution solar and wind by a factor of three compared to fossil fuels'. *EnergyPost.eu*, 31 August. http://energypost.eu/iea-underreports-contribution-solar-wind-factor-three-compared-fossil-fuels/

Scheer, H. (2005) *A Solar Manifesto*. London: Earthscan.

Schleicher-Tappeser, R. (2012) 'How renewables will change electricity markets in the next five years'. *Energy Policy* 48 (September): 64–75. https://www.sciencedirect.com/science/article/pii/S0301421512003473?via%3Dihub

Schumacher, I. (2019a) 'Climate policy must favour mitigation over adaptation'. *Environmental and Resource Economics* 4: 1519–31. https://link.springer.com/epdf/10.1007/s10640-019-00377-0

Schumacher, I. (2019b) 'Reducing emissions is a "public good"'. *The Ecologist*, 4 October. https://theecologist.org/2019/oct/04/reducing-emissions-public-good

Searle, S. and Pavlenko, N. (2019) 'Gas definitions for the European

Union'. The International Council on Clean Transportation. https://ec.europa.eu/info/sites/info/files/icct_-_gas_definitions_ for_the_european_union.pdf

Shaw, J. (2018) 'Shocker: US leading all Paris Accord signatories in emissions reduction'. *Hot Air*, 21 August. https://hotair. com/archives/2018/08/21/shocker-u-s-leading-paris-accord-signatories-emissions-reduction/

Shearer, C. (2019) 'How plans for new coal are changing around the world'. *Carbon Brief*, 13 August. https://www.carbonbrief.org/ guest-post-how-plans-for-new-coal-are-changing-around-the-world

Shearer, C., Mathew-Shah, N., Myllyvirta, L., Yu, A. and Nace, T. (2018) 'Boom and bust 2018 tracking the global coal plant pipeline', Coalswarm/Sierra Club/Greenpeace report. https://endcoal.org/ wp-content/uploads/2018/03/BoomAndBust_2018_r6.pdf

Shell (2018) '*Sky* scenario'. Shell International. https://www.shell. com/energy-and-innovation/the-energy-future/scenarios/shell-scenario-sky.html

Shellenberger, M. (2019) TEDx Danubia talk. http://www.youtube. com/watch?v=N-yALPEpV4w

Shojaeddini, E., Naimoli, S., Ladislaw, S. and Bazilian, M. (2019) 'Oil and gas company strategies regarding the energy transition'. *Progress in Energy* 1(1). https://iopscience.iop.org/ article/10.1088/2516-1083/ab2503

Simon, F. (2017) '"Game over" for CCS, driven out by cheap renewables'. *Euractiv*, 4 December. http:// www.euractiv.com/section/climate-environment/news/ game-over-for-ccs-driven-out-by-cheap-renewable

SkySails (2019) SkySails Group GmbH web site. http://www.skysails. info/en/skysails-marine/skysails-propulsion-for-cargo-ships/

Smil, V. (2016) 'Examining energy transitions: A dozen insights on performance'. *Energy Research & Social Science* 22: 194–7. http:// www.sciencedirect.com/science/article/pii/S2214629616302006

Smil, V. (2019) *Growth: From Microorganisms to Megacities*. Cambridge, MA: MIT Press.

Snowden, S. (2019) 'Solar power stations in space could supply the world with limitless energy'. *Forbes*, 12 March. https://www.forbes.com/sites/scottsnowden/2019/03/12/ solar-power-stations-in-space-could-supply-the-world-with-limitless-energy/#251ec7354386

Sørensen, B. (2014) *Energy Intermittency*. Boca Platon, FL: CRC Press. http://www.routledge.com/books/details/9781466516069/

Sovacool, B. (2013) *Energy and Ethics Justice and the Global Energy Challenge*. Basingstoke: Palgrave.

Sovacool, B. (2016) 'How long will it take? Conceptualizing

the temporal dynamics of energy transitions'. *Energy Research & Social Science* 13: 202–15. https://www.sciencedirect.com/science/article/pii/S2214629615300827

Sovacool, B. (2019) University of Sussex press release, 12 April. https://www.sussex.ac.uk/news/all?id=48423

Sovacool, B. and Walter, G. (2019) 'Internationalizing the political economy of hydroelectricity: Security, development and sustainability in hydropower states'. *Review of International Political Economy* 26(1). https://www.tandfonline.com/doi/full/10.1080/09692290.2018.1511449

Spector, J. (2018) 'Lithium-ion storage installs could grow 55% every year through 2022'. Greentech Media, 22 August. https://www.greentechmedia.com/articles/read/lithium-ion-storage-installations-could-grow-by-55-percent-annually#gs.7lCNGU4

STA (2019) Explainer: Solar Farm: Solar Trade Association. https://www.solar-trade.org.uk/solar-farms/

Stanford (2019) Abstracts of 42 peer-reviewed 100% renewables papers, compiled by Mark Jacobson, Stanford University. http://web.stanford.edu/group/efmh/jacobson/Articles/I/CombiningRenew/100PercentPaperAbstracts.pdf

Stark, C. (2019a) 'Towards net zero'. Committee on Climate Change, 19 March. http://www.theccc.org.uk/2019/03/19/chris-stark-towards-net-zero/

Stark, K. (2019b) 'EIA Outlook 2019: The "extremely conservative" case for renewables growth'. Greentech Media, 1 February. https://www.greentechmedia.com/articles/read/eia-outlook-conservative-renewables#gs.xhhs19

State of Green (2018) 'Renewable energy sources are replacing coal in the Danish energy mix'. State of Green Danish web site, 27 November. https://stateofgreen.com/en/partners/state-of-green/news/renewable-energy-sources-are-replacing-coal-in-the-danish-energy-mix/

Stern, N. (2007) The Economics of Climate Change. Report for UK Treasury. Cambridge: Cambridge University Press. https://www.cambridge.org/us/academic/subjects/earth-and-environmental-science/climatology-and-climate-change/economics-climate-change-stern-review?format=PB

Stevenson, A. (1965) 'Strengthening the international development institutions'. Speech before the United Nations Economic and Social Council, Geneva, Switzerland, 9 July. http://www.adlaitoday.org/articles/connect2_geneva_07-09-65.pdf

Stoker, L. (2018) 'UK solar costs plummeting beyond forecasts, as cheap as £40/MWh by 2030'. *Solar Power Portal*, 12 December. https://www.solarpowerportal.co.uk/news/uk_solar_

costs_plummeting_beyond_forecasts_as_cheap_as_40_mwh_
by_2030

Swift-Hook, D. (2016) 'Renewables are for saving fuel'. Blog
post. http://donald.swift-hook.com/index.php/renewables-are-
for-saving-fuel/

T&E (2018) 'Roadmap to decarbonising European aviation'.
European Federation for Transport and Environment.
https://www.transportenvironment.org/sites/te/files/
publications/2018_10_Aviation_decarbonisation_paper_final.
pdf

Teirstein, Z. (2019) 'Climate movement grandpa James Hansen
says the Green New Deal is "nonsense"'. *Grist*, 24 April. https://
grist.org/article/climate-movement-grandpa-james-hansen-says-
the-green-new-deal-is-nonsense/

Teyssen, J. (2012) 'E.on confirms strategy focused on renewables'. *Windpower Monthly*, quoting E.ON executive
Teyssen's comments to German business daily *Handelsblatt*,
30 March. http://www.windpowermonthly.com/go/windalert/
article/1124937/?DCMP=EMC-CONWindpowerWeekly

Than, K. (2018) 'Critical minerals scarcity could threaten renewable
energy future'. Stanford University. https://earth.stanford.edu/
news/critical-minerals-scarcity-could-threaten-renewable-energy-
future#gs.vF06b5Bj

Thomas, S., Dorfman, P., Morris, S., Ramana, M. V. (2019)
'Prospects for small modular reactors in the UK and worldwide'.
Nuclear Consulting Group, London/Nuclear Free Local
Authorities. https://www.nuclearconsult.com/wp/wp-content/
uploads/2019/07/Prospects-for-SMRs-report-2.pdf

Thurston, C. (2017) 'As Mexican solar auction prices scrape
bottom, will quality be threatened?'. *Renewable Energy
World*, 27 November. https://www.renewableenergyworld.com/
articles/2017/11/as-mexican-solar-auction-prices-scrape-bottom-
will-quality-be-threatened.html

Timperley, J. (2019) 'Renewable hydrogen "already cost competitive", says new research'. *Energypost.eu*, 15 March. https://
energypost.eu/renewable-hydrogen-already-cost-competitive-
says-new-research/

Tirone, J. (2019) 'Nuclear faces climate-fight irrelevance
without lower cost'. Bloomberg, 7 October. https://
www.bloomberg.com/news/articles/2019-10-07/nuclear-
called-irrelevant-in-climate-fight-without-lower-costs

Trainer, T. (1995) *The Conserver Society Books*. London: Zed.

Trainer, T. (2012) 'A critique of Jacobson and Delucchi's
proposals for a world renewable energy supply'. *Energy Policy*

44:    476–81.   http://www.sciencedirect.com/science/article/pii/ S0301421511007269

Tröndle, T., Pfenninger, S. and Lilliestam, J. (2019) 'Home-made or imported: On the possibility for renewable electricity autarky on all scales in Europe'. *Energy Strategy Reviews* 26. https://www. sciencedirect.com/science/article/pii/S2211467X19300811

UNFCCC (2016) 'What is the Paris Agreement?', December 2015. https://unfccc.int/process-and-meetings/the-paris-agreement/ what-is-the-paris-agreement

Vahrenholt, F. (2017) 'Germany's Energiewende: A disaster in the making'. London: Global Warming Policy Foundation. https:// www.thegwpf.org/content/uploads/2017/02/Vahrenholt-20171. pdf

Varoufakis, Y. and Adler, D. (2019) 'It's time for nations to unite around an International Green New Deal'. *The Guardian*, 23 April. https://www.theguardian.com/commentisfree/2019/apr/23/ international-green-new-deal-climate-change-global-response

Vaughan, A. (2018) 'UK electricity use falls – as rest of EU rises'. *The Guardian*, 30 January. https://www.theguardian.com/ business/2018/jan/30/uk-electricity-use-falling-economy-weather

Vaughan, A. (2019) 'Electric cars won't shrink emissions enough – we must cut travel too'. *New Scientist*, 20 March. https://www. newscientist.com/article/2197211-electric-cars-wont-shrink-emissions-enough-we-must-cut-travel-too/

Vivid Economics (2018) 'Thermal generation and electricity system reliability'. Vivid Economics/Imperial College London consultants, report for the Natural Resource Defence Council. http:// www.vivideconomics.com/publications/thermal-generation-and-electricity-system-reliability

Waldman, J., Sharma, S., Afshari, S. and Fekete, B. (2019) 'Solar-power replacement as a solution for hydropower foregone in US dam removals'. *Nature Sustainability* 2: 872–8. https://www. nature.com/articles/s41893-019-0362-7.epdf

Wartsila (2019) 'Towards a 100% renewable energy future'. Wartsila global energy company presentation. www.wartsila. com/docs/default-source/power-plants-documents/downloads/ presentation/towards-a-100-renewable-energy-future---presen tation.pdf

Watts, J. (2019) '"Biggest compliment yet": Greta Thunberg welcomes oil chief's "greatest threat" label'. *The Guardian*, 5 July. https://www.theguardian.com/environment/2019/jul/05/ biggest-compliment-yet-greta-thunberg-welcomes-oil-chiefs-greatest-threat-label

Weaver, J. (2019) '"Unlimited" capacity hydrogen storage

facility developing in Utah'. *PV Magazine*, 3 June. https://pv-magazine-usa.com/2019/06/03/unlimited-capacity-hydrogen-storage-facility-developing-in-utah/

WEC (2011) 'Global transport scenarios 2050'. London: World Energy Council. https://www.worldenergy.org/publications/entry/world-energy-scenarioes-global-transport-scenarios-2050

WEC (2017) 'The developing role of blockchain'. London: World Energy Council. https://www.worldenergy.org/publications/entry/the-developing-role-of-blockchain

WEC (2019) 'World energy insight brief'. London: World Energy Council. https://www.worldenergy.org/publications/entry/innovation-insights-brief-global-energy-scenarios-comparison-review

Wehrmann, B. and Wettengel, J. (2019) 'Polls reveal citizens' support for Energiewende'. *Clean Energy Wire*, 14 February. https://www.cleanenergywire.org/factsheets/polls-reveal-citizens-support-energiewende

Wei, T. and Liu, Y. (2017) 'Estimation of global rebound effect caused by energy efficiency improvement'. *Energy Economics* 66: 27–34. https://www.sciencedirect.com/science/article/pii/S0140988317301949

Weißbach, D., Ruprecht, G., Hukea, A., Czerskia, K., Gottlieba, S. and Husseina, A. (2013) 'Energy intensities, EROIs (energy returned on invested), and energy payback times of electricity generating power plants'. *Energy* 52, 1 April: 210–21. http://www.sciencedirect.com/science/article/pii/S0360544213000492#aff4

Weißbach, D., Ruprecht, G., Hukea, A., Czerskia, K., Gottlieba, S., and Husseina, A. (2014) 'Reply on Comments on "energy intensities, EROIs (energy returned on invested), and energy payback times of electricity generating power plants"'. *Energy* 68, 15 April: 1004–16. http://www.sciencedirect.com/science/article/pii/S0360544214001601

Welfle, A., Gilbert, P. and Thornley, P. (2014) 'Securing a bioenergy future without imports'. *Energy Policy* 68: 1–14. http://www.sciencedirect.com/science/article/pii/S0301421513012093

Weston, P. (2019) 'Tackle climate change by fertilising ocean with iron, expert says'. *The Independent*, 4 July. http://www.independent.co.uk/environment/climate-change-ocean-iron-aerosols-fertilise-science-david-king-a8988241.html

Wettengel, J. (2018) 'Significant drop in energy use pushes down German emissions in 2018'. *Clean Energy Wire*, 19 December. http://www.cleanenergywire.org/news/significant-drop-energy-use-pushes-down-german-emissions-2018

Willis, R. (2019) 'To tackle the climate crisis we need more democracy, not less'. *The Conversation*, 21 June. https://theconversation.com/to-tackle-the-climate-crisis-we-need-more-democracy-not-less-119265

WNISR (2017) 'World nuclear industry status report'. https://www.worldnuclearreport.org/The-World-Nuclear-Industry-Status-Report-2017-HTML.html

WNISR (2019) World nuclear industry status report. https://www.worldnuclearreport.org/

WNN (2017) 'China plans further high temperature reactor innovation'. *World Nuclear News*, 19 September. http://www.world-nuclear-news.org/NN-China-plans-further-high-temperature-reactor-innovation-1909171.html

Wood, G. (2019) 'Managing the decline of fossil fuels: A long goodbye', in G. Wood and K. Baker (eds), *The Palgrave Handbook of Managing Fossil Fuels and Energy Transitions*. Basingstoke: Palgrave Macmillan, pp. 611–16.

WWF (2011) 'Positive energy: How renewable electricity can transform the UK by 2030'. London: World Wide Fund for Nature. http://assets.wwf.org.uk/downloads/positive_energy_final_designed.pdf

WWF (2013) '100% renewable energy by 2050 for India'. World Wild Fund for Nature, with TERI, New Delhi. http://www.wwfindia.org/?10261/100-Renewable-Energy-by-2050-for-India

WWF (2014) 'China's future generation'. World Wide Fund for Nature. http://worldwildlife.org/publications/china-s-future-generation-assessing-the-maximum-potential-for-renewable-power-sources-in-china-to-2050

Wyman, O. (2019) 'As more EVs hit the road, blackouts become likely'. *Forbes*, 15 May. https://www.forbes.com/sites/oliverwyman/2019/05/15/as-more-evs-hit-the-road-blackouts-become-likely/#5fcc97a3dc30

Ye, Q., Dong, W., Dong, C. and Huang, C. (2018) 'Fixing wind curtailment with electric power system reform in China'. Washington, DC: Brookings Institution. http://www.brookings.edu/research/fixing-wind-curtailment-with-electric-power-system-reform-in-china

Young, I. and Ribal, A. (2019) 'Multiplatform evaluation of global trends in wind speed and wave height'. *Science* 364(6440): 548–52. https://science.sciencemag.org/content/364/6440/548

Zappa, W., Junginger, M. and den Broek, M. (2019) 'Is a 100% renewable European power system feasible by 2050?'. *Applied Energy* 233–4: 1027–50: https://www.sciencedirect.com/science/article/pii/S0306261918312790

Index